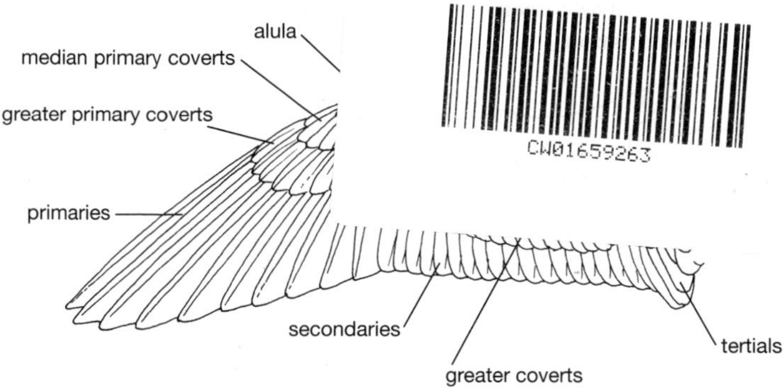

Upperwing (seabird)

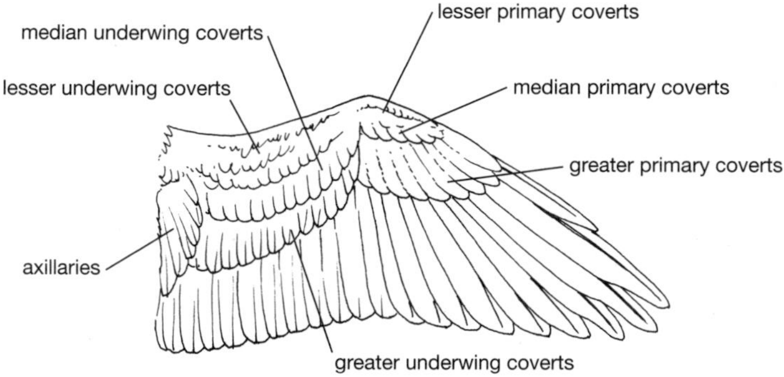

Underwing (passerine)

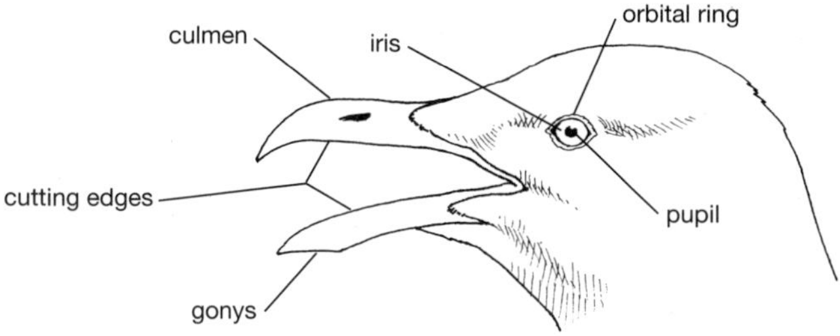

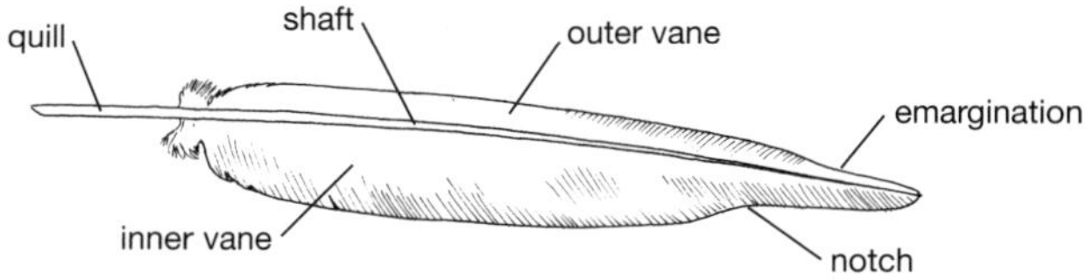

Birds of Spain

Eduardo de Juana
Juan M. Varela

English translation by Ernest Garcia

Edited by:

Lynx

In association with:

SEO BirdLife

Birds of Spain

Third edition in Spanish in 2016:
Guía de las Aves de España. Península, Baleares y Canarias

Texts and maps: ©Eduardo de Juana Aranzana
Illustrations: ©Juan Varela Simó
Cover illustrations: Montagu's Harrier (*Circus pygargus*) and Blue Rock-thrush (*Monticola solitarius*)
Back cover illustrations: Iberian Azure-winged Magpie (*Cyanopica cooki*), Great Bustard (*Otis tarda*), Great Spotted Cuckoo juv. (*Clamator glandarius*), Pin-tailed Sandgrouse (*Pterocles alchata*) and Audouin's Gull (*Larus audouinii*)
Printed in: Ingoprint, S.A.
Depósito Legal: B-1191-2017
ISBN: 978-84-16728-02-2

ÍNDEX

FOREWORD TO THE FIRST EDITION (JULY 2000)

There is undoubtedly a moment in the life of any guide, any book, indeed any work, that is seen to mark its conception. This guide is no exception. For me, its birthplace was a Madrid bar to which we withdrew following a farewell dinner in honour of Pancho Purroy, that great and long-serving Chairman of SEO, the Spanish Ornithological Society.

There were numerous comings and goings among the various small groups of colleagues who were present and there came a moment when two of my closest friends approached me. Here was Juan Varela, that great bird artist who has lent his incomparable style to practically all SEO publications, and numerous others, over so many years. Here too was, Eduardo de Juana, for me the foremost expert on the status of the Spanish avifauna. Both looked somewhat evasive – great secrecy attends the birth of all publishing ventures – before relating their project, at the same time as offering me the opportunity to publish it, via Lynx Edicions.

The idea of a guide to the *Birds of Spain*, Balearics and Canaries included, drawing on the completely homegrown and original information that nobody has mastered as completely as Eduardo – not for nothing do his numerous achievements include leading the Spanish Rarities Committee for over 16 years – and illustrated by the brushwork of Juan, inspired as always by the forms and races of the Spanish avifauna – he is of the old school, jotting in his field notebook everywhere across the length and breadth of the country – seemed most exciting.

Those of us thronging the bar were all SEO members, including half the committee and a good number of the staff. A wider consensus thus soon emerged that such a work should be the 'SEO fieldguide' or, which amounts to the same thing, a useful tool for the conservation of birds and their habitats. It only remained for us to find some feature that would set apart this work from the many other existing guides. We decided on cost; a cover price accessible to all and that would not deter anyone as yet uncertain whether they would be fascinated by the world of birds. A wide readership would stimulate a liking for birding, and an appreciation of nature and the underlying need for its conservation. The SEO Publicity Officer, Fernando Barrio, now became involved, being charged with finding a sympathetic firm that would support the work so as to make the 'reasonable' price possible. He approached *Swarovski* and his search ended there, something for which I wish to express my sincere personal gratitude and that of SEO.

All this has resulted in this book. It is small – and so practical for use in the field – but of impressive quality, content and conservation potential, to my mind a magnificent example of what collaboration between ornithologists and industry, publishers and conservation bodies can achieve. So there you have it; time spent in bars is seldom wasted...

Josep del Hoyo Calduch
Vicechairman of SEO/BirdLife

PREFACE TO THE THIRD EDITION

It has been a long time, 16 years, since this field guide was first published in 2000 and it has been well received, given the 40,000+ sales to date of the Spanish editions. A second comprehensive revision was published in 2005. A further effort has since been made to sustain the interest and usefulness of the guide, resulting in the third Spanish edition in 2016. This last version is now published here in English, translated and edited by Ernest Garcia, in order to make it accessible to a wider readership, many of whom regularly visit Spain for birding.

The guide was intended from the start to support the growing interest in ornithology in Spain, and hence to serve as a contribution to bird conservation there. It was envisaged as a small, straightforward and inexpensive resource for beginners wanting to develop their bird identification skills. We realised that we should only include the birds of Spain. Available guides cover the whole European avifauna, or even that of the entire Western Palearctic, making them larger and costlier, and harder for beginners to use. Dealing with Spain alone has meant that space could be devoted to more detailed information specific to this country, dealing with such aspects as habitat preferences, the timing of migratory movements, the sizes and numerical trends of populations and, in particular, the geographical distribution of species in the region. This approach, supported by the accessible format, numerous plates and up-to-date texts and maps, is what has been most valued by our readers.

Several aspects of this new edition deserve highlighting. Firstly, we have updated both the taxonomic sequence; of orders, families and species, and the nomenclature itself, the scientific names of the species and subspecies currently recognised. The sequence owes a great deal to advances in genetic analyses. Since the pioneering work on DNA hybridisation, by Charles Sibley and his co-workers in the mid 1980s, direct and ever more precise comparisons of both nuclear and mitochondrial DNA have been possible. These have resolved numerous doubts regarding the ancestral parentage of many bird groups in which the situation has been confused by physical similarities due to evolutionary convergence. To give a couple of examples, we now know that the nearest relatives to the flamingos (Phoenicopteriformes) are in fact the grebes (Podicipediformes), despite their very different appearance. Again, the falcons (Falconiformes) are phylogenetically closest to the parrots (Psittaciformes) and passerines (Passeriformes) than to the other diurnal raptors (Accipitriformes), with which they have always been grouped. DNA comparisons are less useful in defining species limits, for which it remains most useful to take differences in coloration and vocalisations into account. However, advances have been made here too, resulting in some former subspecies being recognised as full species. Examples of these 'new' species include the Iberian Green Woodpecker *Picus sharpei*, the Balearic Warbler *Sylvia balearica* and the Iberian Magpie *Cyanopica cooki*. Here we have decided to follow the recent taxonomic proposals put forward jointly by BirdLife International and the *Handbook of the Birds of the World* (del Hoyo, J. & Collar, N. J., 2014 and 2017, *HBW and BirdLife International Illustrated Checklist of the Birds of the World*, vol. 1: Non-passerines; vol 2: Passerines. Lynx Edicions, Barcelona). We realise that the new sequence may be confusing at first but are sure that it will not take too long to get used to it.

The Spanish winter atlas (*Atlas de las aves en invierno en España 2007–2010* de SEO/BirdLife (del Moral *et al.*, 2012) has been especially useful in updating information on bird distributions in winter and the non-breeding period in general. We have also simplified the numerical estimates of breeding bird populations, which remain both a controversial and a highly technical matter. Our treatment of occasional and vagrant species has been updated. Some such species are no longer regarded as rarities in Spain and so have been moved to the main body of the book. Finally, many of the plates have had to be rearranged to accommodate the new species sequence. All the illustrations have been checked for accuracy, and amended where necessary, and several have been replaced by entirely new ones. New illustrations are also included for some of the rare species. We trust that all these changes will be well received.

The authors

INTRODUCTION

This book has a dual purpose. First and foremost it is a fieldguide. Secondly, it offers a brief summary of our knowledge of the Spanish avifauna. It is aimed at a wide readership, specifically those who would like to take up birding and those who are looking for a simple, straightforward and up-to-date reference work on Spanish birds. Experienced ornithologists seeking more detailed information should also consult *The Birds of the Iberian Peninsula* (de Juana, E. & Garcia, E. 2015. Bloomsbury, London).

Two features of this book particularly assist its usefulness in the field: it is small and it is physically resilient. A fieldguide needs to fit a pocket easily and to withstand less than delicate handling for as long as possible. Small size has been largely achieved by limiting the number of species treated to those encountered in Spain alone, although including the birds of the Balearic and Canary Islands and the autonomous North African cities of Ceuta and Melilla. These species though have been given comprehensive treatment. The principal section describes and illustrate all those species that occur regularly at one or other time of year, together with some rarities that occur relatively often. An appendix treats the remaining rarities and vagrants for which there are accepted Spanish records of natural occurrence; 206 species for each of which we summarise the regions of origin, the number of accepted records, and the distribution and timing of occurrences in Spain. Rarity records cover the period up to the end of 2013, as summarised in the most recent available report of the Spanish Rarities Committee, CR/SEO, for 2012/13 (*Ardeola* 63: 453–508).

The principal section covers 372 species, three to five per page (but fewer than four on average). This allows the texts to include all necessary information without resorting to an uncomfortably small font size. The illustrations too are of comfortable dimensions. We have also succeeded in ensuring that the texts and maps are always opposite the illustrations to which they correspond, an essential attribute in a practical fieldguide.

When it comes to identifying birds in the field, knowing when and where to expect a particular species: in which region, habitat or time of year, is almost as important as information on its appearance, voice or behaviour. Restricting this guide to Spain alone has the great advantage that it allows us to include more precise and relevant information. The distribution maps thus offer details on a regional scale. The texts also offer Spain-specific details that would not appear in a Europe-wide guide, including pointers to specific sites where a species is most abundant or easily found. This allows us to meet our second main objective, to provide an introduction to the Spanish avifauna. In addition to notes on status, distribution, relative abundance, ecological preferences and migratory behaviour we have been able to include specific and up-to-date information on population sizes and trends, taxonomy at the subspecific level, winter quarters of summer visitors and conservation categories, among other aspects.

Including all this material has nonetheless required the use of an economical text style and some familiar abbrevations. The text also omits some plumage details that are obvious in the plates and aspects of distribution that are evident in the maps. We have tried throughout to maintain a balance between the different types of information presented, emphasising that which will be most useful for field identification.

The References list gives the most important recent publications on the Spanish avifauna, in particular the ornithological atlases and regional avifaunas, as well as regional Bird Reports, including those published on-line. It also includes other identification guides and those advising on where to watch birds in Spain.

It is worth familiarising yourself with the layout of the species texts, in order to use the guide to maximum effect, especially in the field. The texts comprise several sections separated by a diamond symbol (✦). Each begins with the English name (in colour), scientific name (in italics) and Spanish name (in parentheses), followed by the approximate body length (in bold). The following are then given in order:

1. fieldmarks ✦
2. habitat ✦
3. movements ✦
4. distribution and abundance.

The information given under **fieldmarks** is in order of usefulness in helping to separare the species from other similar ones. It thus often starts with a geographical or seasonal indication that has nothing to do with morphology but indicates where or when that species might be expected (for example: North, Canary Is, winter). For the same reason, descriptions sometimes begin with coloration, where this is distinctive, but sometimes with size, voice or some striking aspect of behaviour. In effect, the information is a sort of key that allows confusion species to be eliminated rapidly in order to arrive at the bird in question. In general, adult plumages are considered before those of juveniles or immatures, and males before females in sexually dimorphic species. The topographical terms used are indicated in both endpapers. Descriptions of vocalisations are mainly limited to those of most help in the field, whether calls or territorial songs, as stated in each case. To save space, characteristics shared by all species in a group may be included in the summary that precedes each family.

The **habitat** section is a brief summary of habitat selection by a species during the season or seasons that it is present in Spain. Altitudinal range is given where relevant together with, where space allows, specific information on nesting habitat.

The **movements** section usually starts with a general statement of the migratory status of the species (e.g. 'trans-Saharan migrant' or 'sedentary, with juvenile dispersion') and ends with summary of the period or periods during which it is regularly found in Spain (e.g. 'October to February'). The latter is a generalisation and space has not always allowed a description of exceptions and variations. This section also indicates regions of origin of migrants and winter visitors, or destinations of summer visitors, again where space allows.

The section on **abundance and distribution** complements two separate distribution map types: one for the Peninsula and Balearic Islands and the other, on a larger scale, for the Canary Islands. The relevant subspecies are indicated, except where the one involved is just the nominate race. Population size is given, usually approximately, where known, together with any information of practical use on distribution or regional variation in abundance (along the lines of 'mainly in Andalucía' or 'rarer in the north'). Known population trends, 'increasing', 'declining', etc., are also given. Specific sites where a species is especially numerous or particularly easy to see are also included where appropriate. The locations of such sites are shown in the maps below. Finally, the Conservation Category within Spain is stated for those species that are included in the Spanish Red Book, the *Libro Rojo de las Aves de España* (Madroño *et al.*, 2004), using the following official abbreviations:

- **CR** = Critically Endangered: the species or subspecies is under extremely high risk of extinction in the wild.
- **EN** = Endangered: under very high risk.
- **VU** = Vulnerable: under high risk.
- **NT** = Near Threatened.
- **DD** = Data Deficient: there is insufficient information to evaluate the risk of extinction adequately, it being possible that the species or subspecies being under threat.

The **Distribution Maps** use different colours to distinguish between ranges in different seasons.

Green = Resident, i.e. present year-round.
Yellow = Breeding, present only in the breeding season.
Blue = Winter, present only in winter. Dark blue = common in winter, light blue = scarce or sporadic in winter.

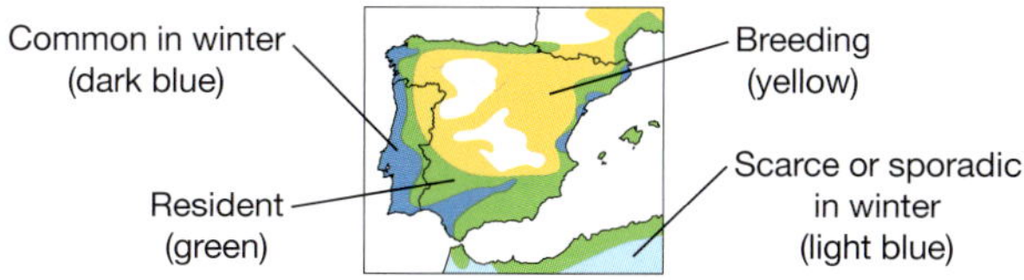

LOCALITIES OF INTEREST MENTIONED IN THE TEXT

1.- Aiguamolls (del Ampurdán), marsh
2.- Alborán, island
3.- Albufera (de Valencia)
4.- Albufera de Mallorca
5.- Alfés, dwarf shrub steppe
6.- Arosa, ría
7.- Belchite, steppes
8.- Benidorm, islet
9.- Cádiz, bay
10.- Cádiz, lagoons
11.- Chafarinas, islands
12.- Cíes and Ons, islands
13.- Columbretes, islands
14.- Daimiel, washes
15.- Delta (Ebro)
16.- Doñana
17.- El Hondo, reservoirs
18.- Estaca de Bares, cape
19.- Strait of Gibraltar
20.- Formentera, saltpans
21.- Fuente de Piedra, lake
22.- Gallocanta, lake
23.- Gasset, reservoir
24.- Gijón, bay
25.- Gredos, sierra
26.- Grosa, island
27.- Guernica, ría
28.- La Mancha, lakes
29.- La Mata, saltpans
30.- La Nava, lake
31.- La Serena, steppes
32.- Llanos de Cáceres, steppes
33.- Mar Menor, coastal lagoon
34.- Marismas (del Guadalquivir)
35.- Marjal del Moro, lake
36.- Monegros, steppes
37.- Navalcán, reservoir
38.- Nervión, ría
39.- Odiel, marismas
40.- Orellana and Sierra Brava, reservoirs
41.- Peñas, cape
42.- Rosas, bay
43.- San Pedro, sierra
44.- Santander, bay
45.- Santa Pola, saltpans
46.- Santoña, marismas
47.- Sisargas, islands
48.- Tierra de Campos, steppes
49.- Ullibarri, reservoir, and Salburua, pools
50.- Villafáfila, steppes and lakes

CANARY ISLANDS

51.- Alegranza, island
52.- Anaga, rocks
53.- Garachico, rock
54.- Garajonay, laurel forest
55.- Montaña Clara, island
56.- Salmor, rock
57.- Teide, peak

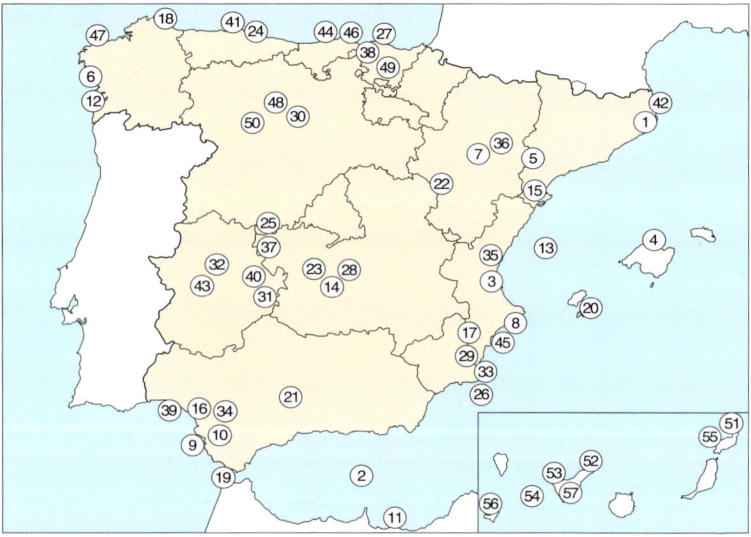

SOURCES

The References section gives the numerous works that have been consulted when compiling this guide. Some especially useful ones deserve particular mention. Regarding identification, we have been greatly helped by Jonsson (1992), Snow and Perrins (1998), Beaman and Madge (1998) and the Collins guide, Svenson *et al.* (2010), this last undoubtedly being the best currently available European fieldguide. Biometric data is based partly on the Collins guide and partly on Snow and Perrins (1998), the figures being rounded up since they are only intended to be an approximate indication of size. Vocalisations are based as far as possible on our own field experience supplemented by listening to recordings; our transcriptions have necessarily been amended by our translator to correspond with English phonetics. Taxonomy and both English and Spanish names follow the *Handbook of the Birds of the World Alive* (del Hoyo *et al.* 2016; http://www.hbw.com/).

The distributions of birds in Spain are based on the atlas works of SEO/BirdLife, the *Atlas de las aves reproductoras de España* (Martí & del Moral, 2003) and *Atlas de las aves en invierno en España 2007*–2010 (del Moral *et al.*, 2012), simplified in view of the small size of our maps. The *Aves del Archipélago Canario* (Martín & Lorenzo, 2001) has greatly helped our treatment of the Canary Islands, complemented by a revision of our text by Juan Antonio Lorenzo Gutiérrez (Delegación de Canarias de SEO/BirdLife). The following sources have been especially useful in determining bird distribution in neighbouring countries, i.e. continental Portugal and parts of France, Morocco and Algeria: Snow and Perrins (1998: *The birds of the Western Palearctic, concise edition*), Hagemeijer and Blair (1997: *The EBCC atlas of European breeding birds*), Catry *et al.* (2010: *Aves de Portugal. Ornitologia do território continental*), Rufino (1989: *Atlas das aves que nidificam en Portugal Continental*), Dubois *et al.* (2008: *Nouvel inventaire des oiseaux de France*), Thévenot *et al.* (2003: *The birds of Morocco*) and Isenmann and Moali (2000: *Oiseaux d'Algerie*). A full and extensive Bibliography of the Iberian avifauna may be consulted in *The birds of the Iberian Peninsula* (de Juana and Garcia, 2015).

GAME BIRDS (Galliformes, Phasianidae) – Partridges, grouse and allies. Terrestrial omnivores that scrabble in soil. Bills strong, short and somewhat recurved; legs sturdy, sometimes spurred, and feathered in grouse. Wings short and rounded; take-off explosively. Males often larger and more brightly coloured. Large clutches laid in ground nests. Chicks precocial, nidifugous. Gregarious; largely sedentary. Much hunted. Occur worldwide.

Rock Ptarmigan *Lagopus muta* (lagópodo alpino). 33–40 cm. Pyrenees. The 'snow partridge', unmistakable in all its seasonal plumages although always well camouflaged. Almost entirely white in winter but with black rectrices (and black lores in males); in spring upperparts become greyish or brownish in patches (males develop large red supraorbital wattles); by late summer largely greyish, white confined to underparts and wings. Scuttles nimbly among rocks and shrubs. Takes flight downhill, partridge-fashion, showing white wings and black tail. Inconspicuous although often quite confiding; has spectacular territorial flights. Flocks in winter ✦ High mountains, >2,000 m, on rocky ground with grassy and dwarf shrub patches ✦ Sedentary, with short altitudinal movements ✦ Ssp. *pyrenaica*: Scarce (a few hundred pairs; numbers fluctuate) and confined to the Pyrenees (mainly in Lleida and Huesca). **VU**

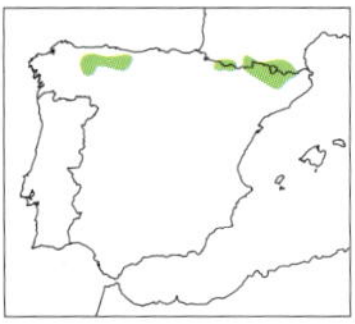

Western Capercaillie *Tetrao urogallus* (urogallo común). 55–90 cm. Northern mountains. A forest bird and hard to see; large droppings reveal its presence. The very large, dark cocks sing and display together at dawn in Apr–Jun at regular locations (leks). Song complex, audible only at close range (an accelerating series of clicks, followed by a sharp '*pop*' note and then a prolonged grinding hiss). The hens are much smaller, highly cryptic, rufous-brown with orange brown throat and breast. Takes flight with very noisy wingbeats; flight alternates wingbeats and glides ✦ Mature and mainly extensive mountain forests, chiefly of conifers in the Pyrenees (at 1,700–2,000 m) and of birches and other deciduous trees in the Cantabrian range (at 1,200–1,600 m) ✦ Sedentary ✦ Ssp. *aquitanicus*: Pyrenees (mainly Lleida and Huesca), scarce (~600 males) and declining; **EN**. Ssp. *cantabricus*: Cantabrian range (Asturias and León), very scarce and rapidly declining (nearly 600 males in 1982, now ~250, almost lost from Cantabria and recently disappeared from Lugo and Palencia). **EN**

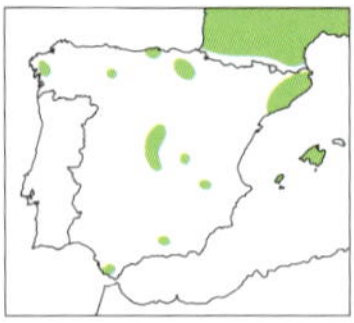

Common Pheasant *Phasianus colchicus* (faisán vulgar). 53–90 cm. Introduced, very local. Much larger than a partridge and long-tailed (except growing chicks). Males very showy; have a loud, harsh crow: *corrr-corrr*, followed by a wing whirr ✦ Selects mixes of copses with meadows and crops, in Atlantic and submediterranean landscapes ✦ Sedentary ✦ Scarce (a few hundred pairs). Numerous releases for hunting have given rise to breeding populations, some apparently stable, especially in Catalonia (Barcelona, Girona); elsewhere generally restricted to enclosed parks and hunting estates. Also introduced to Mallorca and Ibiza.

Rock Ptarmigan

Western Capercaillie

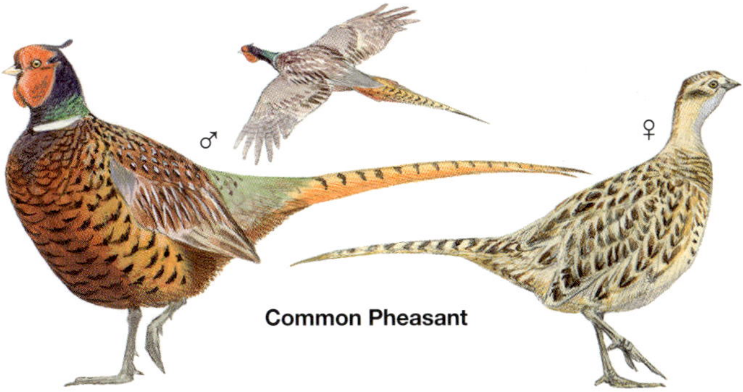

Common Pheasant

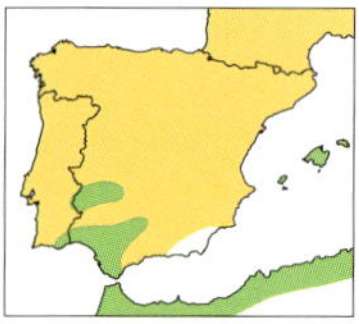

Common Quail *Coturnix coturnix* (codorniz común). 16–20 cm. More often heard than seen, male gives its loud '*quic qui-ick*' song from cover. Small, brown and compact, with pale stripes. Head markings separate it from partridge chicks; male has black throat centre. Looks chunky in flight and soon drops into cover ✦ In irrigated or dry croplands of cereals, legumes etc, and in grasslands, even in mountains ✦ Highly migratory. Summer visitor (Mar/May–Aug/Oct), although some remain in winter (Extremadura, Andalucía, Canary Is). Spring arrivals may stay to breed or continue further, according to seasonal rainfall and plant growth. After nesting many birds, including young of the year, migrate to central Europe amd nest there in summer ✦ Common, locally abundant in some areas and years. Historical decline noted; large interannual fluctuations occur. Much hunted; birds released widely for hunting are often the similar **Japanese Quail** *Coturnix japonica* (codorniz japonesa) or hybrids with that species. Uncommon in Canary Is (ssp. *confisa*), where numbers also fluctuate. **DD**

Red-legged Partridge *Alectoris rufa* (perdiz roja). 34–38 cm. Common. Contrasting head markings; red bill and legs. Sexes similar but males are sturdier with larger heads and have spurs. Young bird has a cream throat and only hints of a collar and flank bars; retains two cream-spotted outer primaries during first year. Wanders rapidly and prone to take flight, with rapid wingbeats and glides on arched wings. Often calls, often from a prominent songpost: *ku-chuk ku-chuk–chuk* or *chuk–chuk–chuk chukuk–chukar*. In pairs in spring, otherwise in flocks ✦ Prefers extensive Mediterranean crop mosaics offering good cover (field margins, fallows, scrub, vineyards, olive groves), but occupies many habitat types at various altitudes (generally <1,500 m) ✦ Sedentary ✦ Two races, much interbred: *hispanica* in N and NW, darker and more brightly coloured, and *intercedens* elsewhere. Very common although declining, locally abundant (e.g. in La Mancha) but rarer in the N. Many are released for hunting. In Canary Is introduced to Gran Canaria (common), and releases occur on other islands. **DD**

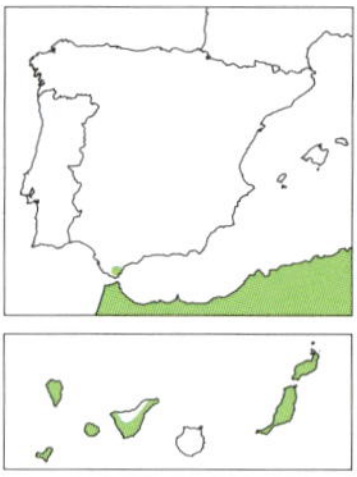

Barbary Partridge *Alectoris barbara* (perdiz moruna). 34–38 cm. Canary Is and Gibraltar. Similar to Red-legged but with face/throat grey, rufous white-spotted collar, dark cap and scapulars marked rufous and bluish. Voice: *ku-chuk*, *ku-chuk*, *ku-chukor, ku-chukor*; also a prolonged grating *krrraaik* ✦ Open terrain: mountainsides, scrub, sparse pinewoods, croplands, semideserts, up to >2,000 m ✦ Sedentary ✦ Ssp. *koenigi*: Canary Is, widespread and common but numbers and densities very variable (abundant in Fuerteventura, S Tenerife and La Gomera; very rare in Gran Canaria); often released. Ssp. *barbara*: introduced to Gibraltar, has been found in Cádiz province and scarce in Ceuta and Melilla.

Grey Partridge *Perdix perdix* (perdiz pardilla). 29–31 cm. Northern mountains. Cryptic, with dull legs and bill, orange-brown face and diffuse flank bars. Horseshoe mark on belly is smaller and less prominent in females. Juveniles as Red-legged Partridge but lack any collar and have white streaks below. In coveys for most of year. Elusive and difficult to surprise; takes flight less noisily than Red-legged, giving alarm call: *kip*, *kip*, *kip*… Song: *skierrrrrrek, skierrrrrrek* ✦ Montane scrub chiefly at ~1.000–2.000 m (moorland; broom, juniper and bearberry scrub). Descends to croplands in winter ✦ Sedentary, moves lower when snow driven ✦ Ssp. *hispaniensis*: scarce (a few thousand), severely declining (Pyrenees, Cantabrian and N Iberian Ranges, Montes de León). **VU**

Common Quail
♀
♂
Red-legged Partridge
Barbary Partridge
Grey Partridge

WILDFOWL (Anseriformes, Anatidae) – Medium-sized or large swimming birds. Bills short and broad, fringed by lamellae originally serving for filtration. Legs short, set well back; webbed feet. Clutches large, young nidifugous. Sociable, often in large concentrations. Generally migratory; many only winter in Spain. Cosmopolitan.

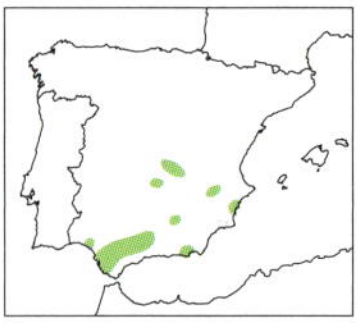

White-headed Duck *Oxyura leucocephala* (malvasía cabeciblanca). 43–48 cm. Stocky and short-necked, with a large head and swollen bill-base (blue in breeding males). Long tail frequently held stiffly upwards. Head variably black and white in males, striped in females and immatures. Short, uniformly coloured wings. Dives frequently but seldom seen in flight. Individuals of the **Ruddy Duck** *Oxyura jamaicensis* (malvasía canela), a related species of American origin, originating from feral British and French populations reach the Peninsula, although occurrences have greatly declined due to culling programmes; this species has a concave profile; the bright rufous males have a black cap extending below the eye; females and immatures more grey-faced, with more diffuse head striping and white undertail coverts ✦ Lakes, reservoirs or gravelpits of some depth and abundant submerged and emergent vegetation ✦ Resident, but performs moult and drought-avoidance movements ✦ Scarce (22 individuals in 1977, but latterly ~2.000–2.500). Chiefly in the S (Cádiz, Sevilla, Córdoba, Almería, Alicante…) and at some wetlands of La Mancha. Reintroduced to Mallorca. **EN**

Mute Swan *Cygnus olor* (cisne vulgar). 125–165 cm. Scarce and local. Large, very long-necked, white or whitish. Can only be confused with the **Whooper Swan** *C. cygnus* (cisne cantor) or the smaller **Tundra Swan** *C. columbianus* (cisne chico) (p. 224), both vagrants with yellow, black-tipped bills. Adults have reddish-orange bill, with swollen black base especially prominent in males. Juveniles and immatures brownish-grey ✦ Ponds, lakes, estuaries ✦ Recently established locally in the north (Santander, Pontevedra, Aiguamolls). Common in captivity. Also a sporadic winter visitor (principally to N coast after cold spells); accidental in Balearics.

Barnacle Goose *Branta leucopsis* (barnacla cariblanca). 58–71 cm. Highly contrasted, with clear delimited strikingly white face (buff in some adults) and black neck and breast. Mantle and upperwing coverts grey, barred black with white feather edging. Short black bill. Legs black. Resembles vagrant **Canada Goose** *B. canadensis* (barnacla canadiense) (p. 225), which is larger with a pale breast and black forehead ✦ Coastal or inland. Often with Greylag Geese ✦ Winter (Nov–Feb). Breeds in the Arctic (Greenland, Spitsbergen, Novaya Zemlya) but populations of feral origin now breed in Europe (Sweden, Low Countries). Usual winter quarters in British Is and around North Sea ✦ Regular in very small numbers, increasing (Marismas, Villafáfila).

Brent Goose *Branta bernicla* (barnacla carinegra). 55–66 cm. Maritime. Only goose with entirely black head. Small, compact, looks largely black with a white stern at a distance. White collar crescent distinctive, lacking at first in juveniles, which have clear white tips to upperwing coverts. Underparts vary according to race (dark-bellied *bernicla* and more contrasted pale-bellied *hrota* shown). Bill and legs black. Long-winged; flies swiftly ✦ Frequents intertidal mudflats in estuaries and other shallow inlets, feeding on eelgrass ✦ Winter visitor (mainly Dec–Feb) from Siberian tundra (*hrota*, from Greenland and Spitsbergen, considered vagrant) ✦ Annual Galicia and Biscay coast, but scarce and irregular; dozens appear in some years. Occasional inland and S; vagrant in Mediterranean and Canary Is.

♀
♂
White-headed Duck

juv.
Mute Swan
ad.

Barnacle Goose
ad.
Brent Goose
ssp. *bernicla*
ssp. *hrota*

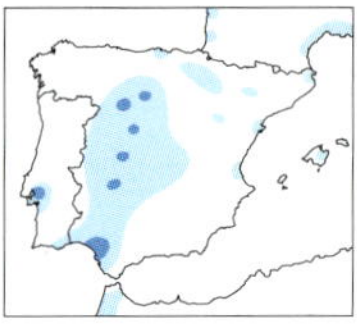

Greylag Goose *Anser anser* (ánsar común). 76–89 cm. The most frequently seen goose and the ancestor of domestic geese. Big and bulky, greyer and less brown than other geese, with a relatively large pale head and large, entirely orange bill. Legs pinkish. Pale grey forewings and rump distinctive in flight. Young birds not obviously barred pale above or on flanks and lack neck 'furrowing'. Strong flier, in V-formation. Voice as farmyard goose; often a loud, cackling *aahng–ang–ang* ✦ Frequents diverse wetlands: marshes, lakes, ricefields, pools, etc. Uproots bulbs of water plants or grazes on pastures, stubble or young crops ✦ Occurs mainly in winter, with passage in Oct/Nov and Feb/Mar (ssp. *anser*, which nests in Denmark, Fennoscandia, Germany, Poland…; some records of pink-billed ssp. *rubrirostris*, from E. Europe and Asia) ✦ Some pairs nest very locally but this trend apparently increasing. Wintering population c. 100,000 birds, showing some recent decline, the great majority in the Marismas, Duero valley lakes (Villafáfila, La Nava, Boada, El Oso…) and Extremadura (Vegas Altas). Scarce in the E and Balearics. Vagrant to Canary Is.

Bean Goose *Anser fabalis* (ánsar campestre). 70–85 cm. Now very rare. Less bulky than **Greylag Goose** and darker brown, especially the long neck and head. Bill dark with a small orange patch. Legs orange. Pale feather fringes of wings and upperparts of adults visible at long range. In flight, wings appear dark and relatively uniform above and below. May be confused at a distance with the **Pink-footed Goose** *Anser brachyrhynchus* (ánsar piquicorto), an occasional winter visitor (p. 225) that has pink legs, grey back and a shorter neck ✦ Forages in young cereal crops ✦ Winter visitor (Nov–Feb) originating from the Russian/Siberian taiga ✦ Ssp. *rossicus*: formerly relatively common around the Esla reservoir (Zamora) but now practically absent since now winters further north in Europe.; **CR**. Vagrant to Balearic Is.

Greater White-fronted Goose *Anser albifrons* (ánsar careto). 65–78 cm. Rare. Adult has prominent white forehead blaze and broad irregular black barring on the belly. May be confused with vagrant **Lesser White-fronted Goose** *Anser erythropus* (ánsar chico) (p. 225), which is much smaller, with yellow orbital ring. First-years lack ventral barring and white blaze begins to develop during the winter; may be confused with **Greylag Goose** but smaller and browner, with orange legs and uniform grey wings, lacking pale forewings. Bill pink in ssp. *albifrons* and orange in *flavirostris* ✦ Marshes and lakes, often accompanying Greylags ✦ Winter visitor (Oct/Nov–Feb/Mar), nesting on tundra in Siberia (*albifrons*) and Greenland (*flavirostris*, vagrant) ✦ Few but increasing records, of single birds or small groups, mainly in the Marismas or Duero valley (Villafáfila, La Nava, etc.). Vagrant to Catalonia, Balearic Is and Canary Is.

ad.
Greylag Goose
ad.
Bean Goose
ad.
Pink-footed Goose
ad.
Greater White-fronted Goose

Long-tailed Duck *Clangula hyemalis* (pato havelda). 37–60 cm (tail included). A small, short-billed duck; white-faced with dark upper neck patches. In flight shows narrow, back-swept, all-dark wings that contrast with pale belly. Dives frequently. Long-tailed males unmistakable although in Spain most are 1st winters and resemble the dark-capped adult females. In winter, adult males show long tail, white scapulars and pink bill patch ✦ Coastal (bays, estuaries, inlets) ✦ Winter (Oct/Nov–Mar/Apr). Nests in Greenland, Iceland, Norway, N. Russia... and winters S to N France ✦ Scattered individuals annual on Biscay coast (Asturias, Santoña) and in Galicia; occasional in Andalucía, Catalonia and the Levant; Vagrant on inland waters and to the Balearic and Canary Is.

Common Eider *Somateria mollissima* (éider común). 50–71 cm. Large, bulky duck. Large head with long conical bill gives triangular profile. Adult male unmistakable but most in Spain are 1st years/immatures, with black bellies, caps and tails and variablly white elsewhere. Females dark brown or greyish, barred. Flight powerful at low levels, often in line formations, in which pied males conspicuous ✦ Shallow coasts ✦ Generally in winter (Sept–Mar, esp. Nov–Jan), but also, rarely, in other months. Range expanding S, now nests in France ✦ Formerly vagrant, individuals and small groups now annual but in very variable numbers, in Catalonia and on E Biscay coast (Basque Country, Santoña, Gijón), occasionally reaching Valencia and Galicia. Vagrant to Balearic Is.

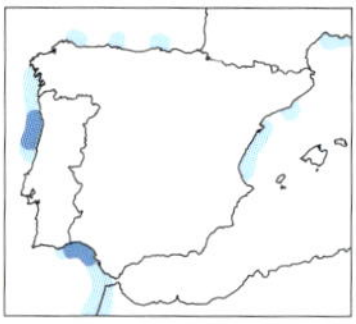

Common Scoter *Melanitta nigra* (negrón común). 43–54 cm. Marine. Males black, with small yellow bill patch. Females dark brown, with paler cheeks contrasting with dark crown (recalling female Red-crested Pochard, p. 24). Compact; bill base swollen; tail pointed; greyish primaries show paler in flight. **Black Scoter** *Melanitta americana* (negrón americano), vagrant in N (p. 225) very similar but male shows extensive yellow on bill. Flies at low levels, in large flocks that often trail into lines ✦ Shallow coasts (bays, beaches) ✦ On passage and in winter (Sept/Nov–Mar/Apr); some records in other months. Breeds N Europe and Iceland, wintering S to Portugal and Morocco ✦ Ssp. *nigra*: thousands on autumn passage on Biscay coasts and off Galicia (La Vaca point, Cape Bares, Cape Silleiro). Common on Atlantic coasts in winter (Huelva, S Galicia, Santoña) but scarce in Mediterranean (Delta, Valencia); probably declining. Vagrant to Balearic and Canary Is.

Velvet Scoter *Melanitta fusca* (negrón especulado). 51–58 cm. Resembles Common Scoter but has striking white secondaries; larger and heavier, with conical bill and reddish legs. Adult males show white mark below eye; females show two, variable pale face patches ✦ Coastal waters ✦ Winter, esp. during cold snaps (most Nov–Feb) ✦ Annual but scarce and declining, in N (Santoña, Santander, Gijón, Rías Bajas) and Catalonia (Roses Bay); rare and irregular in Valencia; occasional in Balearics and Andalucía.

Long-tailed Duck
♀ winter
♂ winter
♂ eclipse
Common Eider
♀
♂
♀
Common Scoter
♂
♀
Velvet Scoter
♂

Common Goldeneye *Bucephala clangula* (porrón osculado). 42–50 cm. Adults distinctive (males show large, white loral spot; females have white collar) but most that reach Spain are less-striking 1st winters (males somewhat larger than females, with white on scapulars). Short-billed, with large, triangular head. Dives actively. Shows white secondaries and upperwing coverts in flight ✦ Salt and fresh waters (estuaries, reservoirs) ✦ Winter (Nov–Mar, most Dec–Feb). Nests N Europe, wintering commonly S to France ✦ Very scarce and irregular, in N (Galicia, Biscay coastlands, Catalonia), more frequent with cold snaps, probably declining. Exceptional or vagrant inland and in Andalucía, Murcia, Valencia and Balearics.

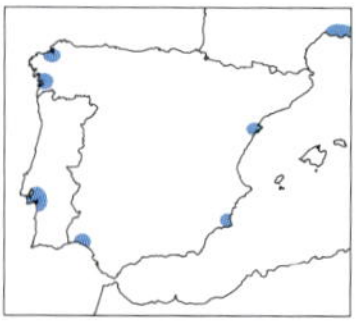

Red-breasted Merganser *Mergus serrator* (serreta mediana). 52–58 cm. Bill very long and slender; shaggy nape crest. Green-headed ad male has white collar but most in Spain are females/immatures, with chestnut heads. Much white on inner wings. Confusable with vagrant **Goosander** *Mergus merganser* (serreta grande) (p. 225); larger, males white-breasted and females/immatures showing sharp, not diffuse, border between chestnut head and grey neck. The much smaller **Smew** *Mergellus albellus* (serreta chica) also occurs rarely in winter (p. 225) ✦ Fishes in sheltered coastal waters (estuaries, inlets, lagoons) ✦ In winter and on passage (Oct/Nov–Mar/Apr). Nests N Europe and Iceland ✦ Rare (a few hundred), local and declining: Rías Bajas, Mar Menor, Huelva estuaries, Delta. On passage on Biscay coasts. Rare in Balearics. Vagrant inland and Canary Is.

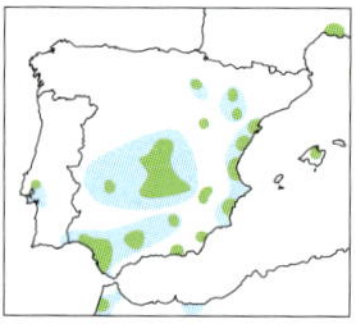

Red-crested Pochard *Netta rufina* (pato colorado). 53–58 cm. Large and showy. Red-billed male, with large rufous head and golden erectile crest, shows sharp contrast between black 'stem and stern' and gleaming white flanks. Female unmarked brownish with characteristic two-tone head and face and pinkish spot near bill tip. Eclipse male as female but bill red. Broad white wing bar striking in flight, as are black belly and white flanks of males. Often dabbles but also dives for food ✦ Nests where there is abundant subaquatic and emergent vegetation. Moults and winters on large waterbodies ✦ Resident, with moult and drought-avoidance movements; joined by C European birds in winter ✦ Population highly variable according to water levels (a few thousand pairs), quite concentrated (Delta, Valencia, La Mancha, Marismas). Moult gatherings occur (latterly esp. at Orellana reservoir, Extremadura). ~15.000 inds. in winter, declining. Introduced to Balearics. **VU**

Common Goldeneye

Red-breasted Merganser

Red-crested Pochard

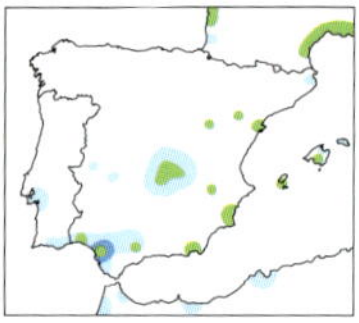

Common Shelduck *Tadorna tadorna* (tarro blanco). 58–67 cm. Boldly-patterned, goose-like duck; unmistakable. Males somewhat larger, with swollen bill base and broader breastband. Juveniles comparatively drab, lacking breastband; white-faced, bill pink ✦ Saltpans, shallow lagoons and pools; in winter and on passage also at marshes, estuaries and on coasts. Selects salt or brackish waters. Nests in burrows in dunes and sandbanks or among rocks ✦ Year-round ✦ Small breeding population (tens of pairs), established in 1970s; now increasing and spreading (Delta, saltpans in Murcia and Alicante, steppe lakes of Ebro valley and La Mancha, Extremadura reservoirs...); has bred Balearics (Ibiza, Formentera, Mallorca); **NT**. Winterers from elsewhere in Europe also occur (a few thousand), increasing, chiefly in the Marismas. Vagrant to Canary Is.

Ruddy Shelduck *Tadorna ferruginea* (tarro canelo). 61–71 cm. Scarce. Fairly uniform rufous coloration unique among European wildfowl. Males more uniformly coloured on head and have black neck collar in summer; females whiter-faced, lack collar. Juveniles browner, with greyer heads. Legs blackish. Large white wing patches striking in flight ✦ Estuaries, lakes, marshes; nests chiefly at shallow, brackish, inland lagoons ✦ Year-round ✦ Birds apparently originating from N Africa formerly wintered regularly at the Marismas (Aug–Mar); more recent observations, chiefly in late summer, suggest visitors from other European countries, where there are feral populations; latterly seems to be becoming established as a breeding species in the Peninsula, with nesting records especially in C, S and W. Vagrant to Balearic Is Recently established in Canary Is (esp. Fuerteventura). **CR**

Egyptian Goose *Alopochen aegyptiaca* (ganso del Nilo). 63-73 cm. Few observations. Somewhat resembles Ruddy Shelduck, with similar white wing patches, but larger, greyer and with dark head and lower breast markings. These last lacking in juveniles, which are duller. Legs reddish, bill pink. Sexes similar ✦ Reservoirs, lakes and other wetlands ✦ Year-round ✦ African species introduced to W Europe (esp. England); increasingly found in Iberia where breeding population gradually becoming established (most records from Catalonia, Extremadura and Andalucía).

Mandarin Duck *Aix galericulata* (pato mandarín). 41–51 cm. Sporadic. Small, compact. Exotic-looking males unmistakable; females/juveniles greyish brown with spotted flanks and white head markings ('spectacles'), bill-base and throat; these last most evident in adult females. Confusable with escaped females/juveniles of **Wood Duck** *Aix sponsa* (pato joyuyo), N. American but frequent in captivity, which has thin flank spots and dark bill nail ✦ Lakes, small reservoirs, river banks; generally with wooded margins ✦ Largely sedentary in Europe but most records in Spain in winter (Nov–Jan) ✦ E Asian but introduced to some European countries, esp. UK. Sporadic. Has nested in Spain and there are records of escaped birds, but observations in general, increasing, suggest winter arrivals from the north (concentrated in Catalonia, Asturias and Galicia).

♂ ad.
Common Shelduck
♂ ad.
Ruddy Shelduck
Egyptian Goose
♂
Mandarin Duck
♀

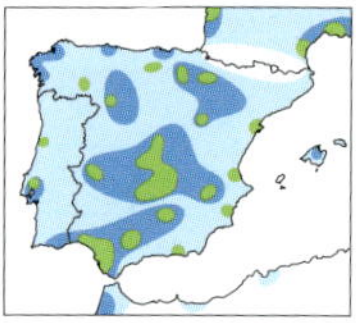

Common Pochard *Aythya ferina* (porrón europeo). 42–49 cm. Common. Male pale grey with black stem and stern, chestnut head and black bill with large blue-grey band. Female greyish-brown; crown, neck and breast darker brown, with pale head markings and smaller blue-grey bill band. Triangular head with sloping forehead and long wide bill; wings grey, with indistinct wingbar. Typical diving duck; heavy, short-necked and short-winged, with whirring flight and tendency to take off and land clumsily ✦ Selects shallow freshwater wetlands: marshes, lakes, gravel pits; at reservoirs in winter ✦ Year-round, with numerous European immigrants in winter (Sept–Apr) ✦ Breeding population highly variable according to water levels, from hundreds to a few thousand pairs, apparently increasing (Marismas, Daimiel...). Common and widespread in winter (some tens of thousands, declining, with concentrations at Gallocanta, El Hondo, Doñana, Delta). Scarce and irregular in winter in Canary Is.

Ferruginous Duck *Aythya nyroca* (porrón pardo). 38–42 cm. Scarce. Very dark, with striking, sharply-defined white undertail patch (unlike female Tufted Duck). Male bright chestnut with white iris; female duller and brown-eyed; juveniles have duller white undertails. Longish necked with long sloping forehead. White wingbar and belly striking in flight ✦ Generally on small freshwater lakes with abundant emergent and submerged vegetation; also on marshes, reservoirs and gravelpits ✦ Year-round, with more records in winter (Oct–Mar) ✦ Hundreds nested formerly in the Marismas but now only breeds there sporadically; scarce breeding records chiefly from the S (Andalucian lakes) and E (Valencia); scarce winterers also in E. **CR**. Beware released or escaped birds. Rare in Balearics and vagrant to Canary Is.

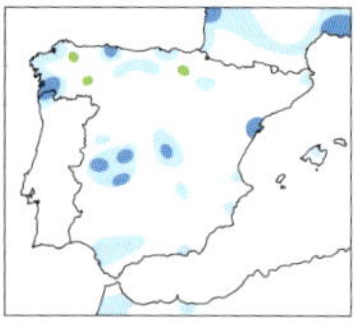

Tufted Duck *Aythya fuligula* (porrón moñudo). 40–47 cm. Male unmistakable; black with white flanks, long hindcrown tuft (short in eclipse and 1st winter birds), bluish bill and yellow eyes. Female dark brown with paler flanks, shorter head tuft, grey-blue bill and yellow eyes; often with white around bill base. Juvenile similar to female, with darker flanks. White wingbars prominent in flight ✦ At relatively deep and open inland and coastal wetlands (lakes, reservoirs, estuaries) ✦ Mainly winter visitor (Oct–Mar, esp. Dec–Feb), from breeding range in C and N Europe ✦ Widespread, but uncommon (few thousand, declining; Galicia, Biscay coastlands, Extremadura, La Mancha, Ebro Delta). A few pairs have nested in recent years across N half of Spain (Salburua, Villafáfila…). Regular winterer in Balearics, scarce and irregular in Canary Is.

Greater Scaup *Aythya marila* (porrón bastardo). 42–51 cm. Scarce. Spanish name based on resembling supposed 'bastard' offspring of Common Pochard (grey-backed) and Tufted Duck (black head and white wingbar). Female shows typical broad and clearly delimited white feathering at bill base; much more evident than in some female Tufted Ducks; in summer most develop a prominent pale auricular patch with wear ✦ Coastal (estuaries, inlets, bays) ✦ Winter visitor (Oct–Apr, most Nov–Feb), from Iceland and N Europe ✦ Very scarce and irregular but recorded annually, most often with cold snaps, in NW (Galicia, Asturias) and NE (Catalonia, Valencia). Occasional in S and inland. Vagrant to Balearic and Canary Is.

Common Pochard
♀
♂
Ferruginous Duck
♂
♀
♂
♀
♂
♀
Tufted Duck
♂
♀
Greater Scaup

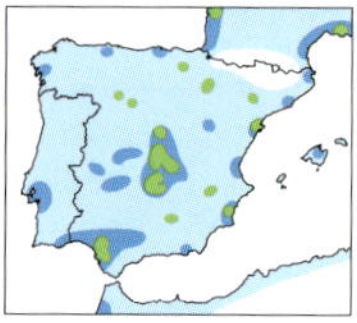

Northern Shoveler *Spatula clypeata* (cuchara común). 44–52 cm. Bill strikingly long. Male unmistakable; dark head, white underparts with chestnut flanks, eye yellow. Female brown but long bill distinctive. Eclipse male more similar to female but more rufous, with pale eye; some autumn males show pale facial crescent ahead of eye. Juvenile resembles darker female. Forehead flat, neck short, speculum green, blue forewing panels are diagnostic, legs orange ✦ Shallow freshwater bodies with marginal vegetation (lakes, gravelpits, irrigation pools); in winter also on marshes, reservoirs and, rarely, on coasts. Bill filters surface water layers ✦ In winter and on passage, from C, N & E Europe (Sept–Apr, most Dec–Feb) ✦ Small, variable breeding population (tens or a few hundred pairs), apparently increasing; **NT**. Abundant in winter (tens of thousands, increasing; Marismas, Sierra Brava, Albufera, Delta, El Hondo). Scarce in Biscay coastlands and Balearics; rare and irregular in Canary Is.

Garganey *Spatula querquedula* (cerceta carretona). 37–41 cm. Somewhat larger, more slender and longer billed than Common Teal. Males have large white crescent above and behind eye, drooping scapulars and dark breast contrasting with pale grey flanks; shows grey-blue forewing and green, white-bordered speculum in flight. Females differ from Common Teal females in having striped head pattern, pale loral spot and no pale stern patch; speculum inconspicuous. Eclipse males, on autumn passage, similar to females but retain pale forewing. Juvenile like female but darker and more streaked below. Male has harsh rattling display call: *rrar ... rrar ... rrar* ✦ Shallow well-vegetated freshwaters ✦ On passage (esp. Mar–Apr and Aug–Sept) between C Europe and tropical W Africa ✦ Common on spring passage, mainly on Mediterranean coast; much rarer in autumn and very rare at other times. Small numbers nest irregularly (tens of pairs; esp. La Nava and Marismas); **VU**. Rare in Canary Is.

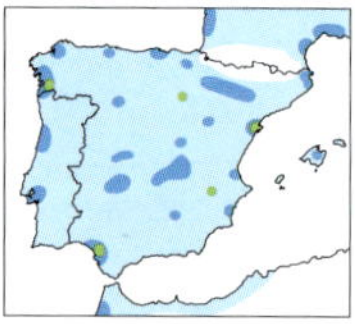

Common Teal *Anas crecca* (cerceta común). 34–38 cm. Small, compact and short-billed. Male: head dark, body grey, shows yellow stern triangle and horizontal white flank stripe (males of ssp. *carolinensis*, vagrant, show instead a vertical white stripe on breast side). Female: rather plain, with white tail sides; often shows yellow at bill base. Juvenile as female but darker and more streaked. Speculum green, shows broad white wingbar across central upperwing. Flies rapidly, often in dense flocks that wheel in unison. Male has distinctive bisyllabic whistle: *prip prip* ✦ Nests at freshwater sites, often small ones and with dense marginal vegetation (ponds, streams); very widespread in winter (marshes, water meadows, ricefields, estuaries, coastal lagoons ✦ Winter visitor (Sept–Apr, most Jan–Feb) from C and N Europe. N American ssp. *A. c. carolinensis* occasional in winter ✦ A few pairs nest in Galicia (Louro river), Asturias (La Ercina lake) and sporadically elsewhere; **VU**. Very common in winter (many tens of thousands, fluctuating; Marismas, Delta, Sierra Brava…). Regular in Balearic and Canary Is.

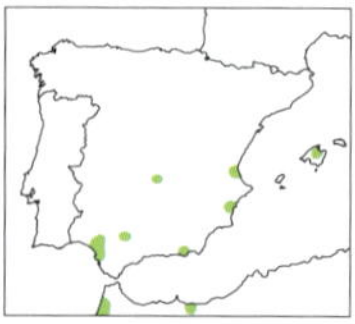

Marbled Teal *Marmaronetta angustirostris* (cerceta pardilla). 39–42 cm. Pale brown overall with whitish blotches and dark eye patch. No speculum. Slim, with long neck and bill. Sexes similar; male shows small nuchal crest and blacker bill ✦ Shallow lakes and pools both freshwater and brackish, in Mediterranean climatic region ✦ Nests May–Jul. Nomadic, avoiding droughts; winter concentrations in S and Morocco ✦ Once abundant (Marismas) but now quite rare and numbers fluctuate widely according to water levels; tens of pairs, most at El Hondo/Santa Pola and the Marismas; forms postnuptial gatherings (El Hondo, >500); **CR**. Occasional in Balearic and Canary Is (has bred Mallorca and Fuerteventura).

Northern Shoveler
Garganey
Common Teal
Marbled Teal

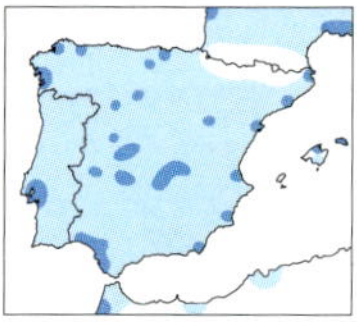

Eurasian Wigeon *Mareca penelope* (silbón europeo). 45–50 cm. Compact, with rounded head, steep forehead, short neck and small bill; pointed stern. Bill bluish-grey, legs grey. Speculum green. Underwing greyish. Males show distinctive golden forehead, pinkish breast, black and white stern; large white upper forewing patch striking in flight and forming small horizontal upper flank line at rest. Females variable dark, with well-defined white belly patch. Eclipse males recall females but more rufous. Flies rapidly on long, narrow wings. Males whistle frequently: *wee–oo* ✦ Coastal wetlands (marshes, lagoons, estuaries) and more seldom inland (reservoirs); grazes ✦ Winter visitor (Oct–Mar) nesting in far northern Eurasia ✦ Common in winter (tens of thousands, declining) especially at the Marismas and other coastal wetlands (Delta, Santoña, Galician rías, Cádiz Bay), also locally inland (Extremadura). Scarce but regular in winter in Balearic and Canary Is.

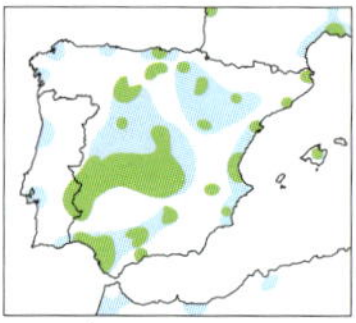

Gadwall *Mareca strepera* (ánade friso). 45–56 cm. Unshowy, with white speculum. Males have elegant vermiculated plumage but appear nearly uniform grey at a distance; with black stern but lacking white on rear flanks. Females recall Mallard females but head and bill more compact, greyer with white belly, yellowish legs and broad orange bill sides. Juveniles similar to females but head greyer and belly spotted. Voice harsher than Mallard ✦ Shallow freshwaters with fringing emergent vegetation (lakes, marshes, reservoirs, gravelpits, pools) where tends to associate with coots and pochards ✦ Resident, with immigrants present in winter (mainly Dec–Feb) ✦ Breeding population small (a few thousand) and variable but tending to increase (Andalucía, La Mancha, Madrid, Extremadura). Quite widespread and common in winter (Marismas, Delta, Sierra Brava). Scarce in Balearics. Scarce and irregular in Canary Is.

Mallard *Anas platyrhynchos* (ánade azulón). 50–65 cm. The common and familiar ancestor of domestic ducks, with which it often interbreeds (beware 'strange' plumages). Largish and strong-billed. Speculum purplish-blue, bordered black and white. At a distance males show darkish head and breast, pale body and black and white stern. Females lack white on belly, have orange legs and bill irregularly blotched black. Eclipse males have yellow bill and reddish-brown breast. Juveniles have greyish bill and streaked breast and flanks. Females give familar, repeated *quack*, males have quiet *rahb* and whistling calls ✦ All wetlands (including reservoirs, park lakes, rivers, streams…); forages on ricefields, young crops, pastures, dehesas ✦ Nests mainly Mar–Apr. Native birds sedentary but dispersive; relatively few immigrants in winter, more in very cold years ✦ Very widespread and common year-round (many tens of thousands), increasing, with local fluctuations according to rainfall. Scarce and irregular in winter in Canary Is.

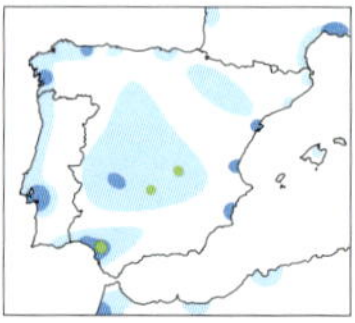

Northern Pintail *Anas acuta* (ánade rabudo). 51–66 cm. Slim, with long slender neck and long tail, especially in males. Bill and legs grey. Males have breast and neck sides white, stern cream and black. Small-headed, attenuated silhouette distinctive in flight ✦ Mainly at coastal wetlands. Selects extensive flooded areas ✦ Winter visitor (most Nov–Feb), from N Europe ✦ A very few pairs nest irregularly, in wet years (esp. Marismas); **VU**. Common in littoral regions in winter, in variable numbers (up to 20,000), most in the Marismas, Albufera, Delta, Galicia); scarce inland (but now numerous in Extremadura) and in Balearics; rare in Canary Is.

Eurasian Wigeon
♀
♂
Gadwall
♀
♂
Mallard
♀
♂
Northern Pintail
♀
♂

GREBES (Podicipediformes, Podicipedidae) – Lobe-footed diving birds of various sizes. Related to flamingos. Bill short and straight. Tail rudimentary. Seldom seen in flight, when show extended neck, projecting feet and short, rapidly-beating wings.

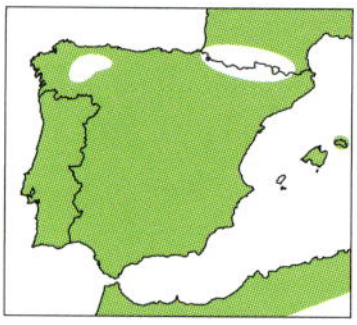

Little Grebe *Tachybaptus ruficollis* (zampullín común). 25–29 cm. Small and dumpy, resembling a toy duck; bill short and straight; stern fluffy. Dark when in breeding plumage, when yellow gape flanges conspicuous; pale brown in winter. Wings lack white. Juvenile has striped head. Call distinctive: a prolonged, high-pitched, whinnying trill. Often solitary but forms groups occasionally ✦ Lakes and pools, often very shallow ones, and riverine backwaters. Prefers abundant emergent vegetation, where it hides. Also on reservoirs, marshes and large water bodies, chiefly in winter. Seldom on the sea ✦ Resident but dispersive ✦ Widespread and common (some thousands of pairs) but scarce in montane or very dry areas. Vagrant to Canary Is.

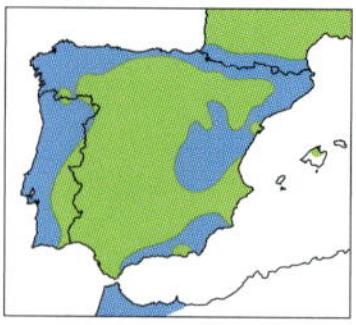

Great Crested Grebe *Podiceps cristatus* (somormujo lavanco). 46–51 cm. Duck-sized and very slender, with unmistakable nuptial head plumes and lappets. In winter long and frequently extended neck appears very white; head white, with dark cap and loral line; bill pale. Juveniles retain striped head into early winter. Courtship striking. Diverse calls in breeding season, often harsh or guttural ✦ Lakes, deep lagoons, reservoirs, gravelpits and irrigation pools, particularly those with abundant fish and emergent vegetation for nesting. More frequent on reservoirs in winter, when hundreds may gather ✦ Year-round. Dispersive, and joined by some immigrants in winter ✦ A few thousand pairs, increasing. Range broad but discontinuous; rare in NW and Biscay coastlands. Scarce winterer in Balearics; has bred Mallorca and Menorca. Vagrant to Canary Is.

Red-necked Grebe *Podiceps grisegena* (somormujo cuellirrojo). 40–50 cm. Only in N, in winter. Large, recalls Great Crested Grebe but more compact, with thicker, grey neck; shorter yellow-based bill and dark cap descending to eyes. Usually solitary ✦ On the sea in sheltered waters (bays, inlets, harbours); occasionally on reservoirs or littoral lagoons ✦ Winters (Nov–Feb). Nearest breeding areas in N Germany and Poland ✦ Ssp. *grisegena*: regular but very scarce, in Biscay coastlands, chiefly Asturias; rarer in Galicia and Catalonia; occasional elsewhere. Vagrant to Balearic Is Ssp. *holboelii* (N America): vagrant.

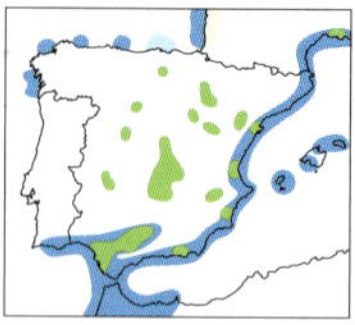

Black-necked Grebe *Podiceps nigricollis* (zampullín cuellinegro). 28–34 cm. Larger and slimmer than Little Grebe, with longer, straighter neck. Breeding plumage showy; black and white in winter, with very white face. Red eye visible at close range. Juveniles show buffish cast on head and neck. Confusable in winter with much rarer **Slavonian Grebe** *Podiceps auritus* (zampullín cuellirrojo) latterly regular in winter in N (Santoña), which is flat-crowned, has a triangular head, a stouter, pale-tipped bill, clean-cut black and white head and neck markings and a sloping stern. Calls diverse. Considerably or very gregarious ✦ Well-vegetated marshes and lagoons, sometimes brackish; in winter also on tidal inlets, bays, saltpans and reservoirs. Nests colonially, on floating vegetation ✦ Resident but many travel some distance and many arrive to winter ✦ Breeding population small and fluctuates according to water levels (a few hundred pairs, in Andalucía, El Hondo, La Mancha lagoons, Gallocanta…). Commoner and more widespread in other seasons, when hundreds or thousands may gather, especially in autumn (Odiel, La Mata, Formentera…). Winters in Balearics and, sparsely and irregularly, in Canary Is. **NT**

Little Grebe
breeding
winter
Great Crested Grebe
winter
breeding
Red-necked Grebe
winter
Black-necked Grebe
winter
breeding
Slavonian Grebe
breeding
winter
transitional

FLAMINGOS (Phoenicopteriformes, Phoenicopteridae) – Large wading birds with extremely long necks and legs and bulky, downward-bent filter-feeding bills. Related to grebes. Plumages pinkish and broad wings bright rufous. Sexes similar. Highly gregarious, nesting in colonies. Make conical nests of mud. Six species, in warm regions of Eurasia, Africa and America.

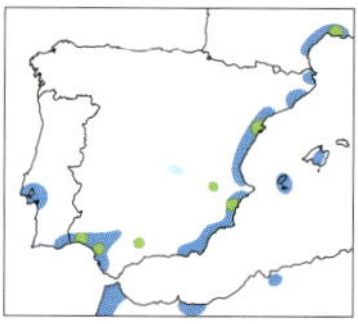

Greater Flamingo *Phoenicopterus roseus* (flamenco común). 125–145 cm. Exotic looking, with extremely long neck and legs. Males larger; young greyish-brown and relatively short-legged; transitional plumages occur ✦ Open shallow waters, both brackish and saline, in saltpans, marshes, bays or lakes. Filters mud ✦ Year-round. Shows long-distance nomadism (Camargue, Tunisia, Sardinia, Morocco…) ✦ Mainly S and E littoral; uncommon in La Mancha and rare or occasional elsewhere. Increasing but population highly variable according to water levels both in winter (4,000–60,000 birds) and breeding season. Most regular colony is Fuente de Piedra, Málaga (>20,000 pairs some years), but has bred in the Marismas and latterly at the Delta, El Hondo and Pétrola (Albacete), with attempts elsewhere. Occasional in Canary Is. **NT**

Lesser Flamingo *Phoeniconaias minor* (flamenco enano). 80–90 cm. Smaller and more rufous than Greater Flamingo; bill dark red. Juveniles brown with dark head and blackish bill and legs ✦ Lakes and saltpans ✦ Year-round (peaks Mar–Jun) in very small numbers ✦ Occasional, from Africa. Most records at Fuente de Piedra, Delta and La Mancha; has bred. Vagrant to Canary Is.

PIGEONS (Columbiformes, Columbidae) – Chiefly arboreal, medium-sized, consumers of leaves, seeds or fruits. Bill short, with cere; head small, legs short. Sexes usually similar. Nests in trees or on rocky ledges. Clutch 1 or 2 eggs, chicks nidicolous. Occur worldwide.

Rock Dove *Columba livia* (paloma bravía). 30–35 cm. Ancestor of domestic pigeons. Upperwings with twin wingbars, underwings silvery. Upper rump conspicuously white (except in Canary Is). Back paler than head; neck has metallic sidepatches, brightest in males but lacking in juveniles; bill dark, iris reddish. Some domestic pigeons identical but others have varied and mixed colouring. Display flight with slow wingbeats, loud wingclapping and glides with wings in a V. Coos like domestic pigeon ✦ Small colonies on sea cliffs and in gorges. Visits cultivation ✦ Sedentary ✦ Wild populations very hard to distinguish from feral ones given frequent hybridisation; the former are uncommon, most frequent in coastal regions and E and S; abundant in Balearics. Ssp. *canariensis*: very common and widespread in Canary Is but declining.

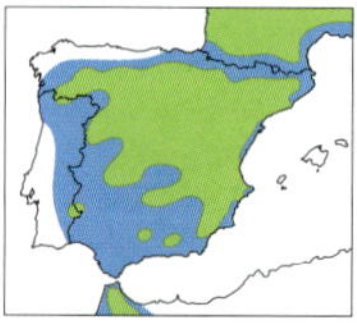

Stock Dove *Columba oenas* (paloma zurita). ~30 cm. Slightly smaller and more delicate than Rock Dove; uniform grey with no obvious white markings. Shows two short black wingbars above; underwings greyish. Round-headed and gentle looking with large dark eyes and yellowish or reddish bill. Wings relatively short and triangular, in flight showing contrasting paler centres. Adults have pinkish breast and iridiscent green neck sidepatches; juveniles much duller. Call: a deep *ooo–uh* repeated in rapid succession. Often in mixed flocks with Woodpigeons ✦ Habitat varied (riparian woodland, dehesas, parks, gorges, steppe areas) but requires holes for nesting, in trees, cliffs or buildings ✦ Partly migratory. Spanish birds perhaps sedentary or dispersive but there is moderate-scale passage and wintering by C European birds (Oct/Nov–Feb/Mar), declining ✦ Quite common (tens of thousands of pairs) but absent from large parts of N, W and S. Vagrant to Balearic Is. **DD**

Greater Flamingo
ad.
ad.
Lesser Flamingo
Rock Dove
Stock Dove

Common Woodpigeon *Columba palumbus* (paloma torcaz). ~40 cm. Conspicuous white transverse band across upperwings (showing at wing bend only at rest) and white neckside patches. Larger, longer-winged and longer-tailed than other pigeons, full breasted with relatively small head. Bill yellow; iris pale, broad black tailband. Wings noisy on take-off. Display flights with steep climbs and wing clapping followed by gliding descent. Song has five syllables: *coo–cooo–coo, coo–coo*, repeated. Often in large flocks ✦ Woodlands, parks, dehesas, orchards. Nests in trees. Frequents crops ✦ Partial migrant; indigenous birds mainly sedentary but many C European birds occur in winter and on passage (Oct–Nov and Feb–Mar) ✦ Breeding population widespread, very common (hundreds of thousands of pairs) and rapidly increasing; ever more abundant in cities. Millions of immigrants, though tending towards decline, arrive mainly across Navarran Pyrenees and cross the passes of the Iberian and Central ranges to winter especially in the dehesas of Extremadura and S Portugal, there sometimes forming immense roosts. Great numbers are hunted. Very common in Balearic Is. Vagrant to Canary Is.

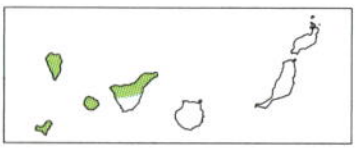

Dark-tailed Laurel-pigeon *Columba bollii* (paloma turqué). ~37 cm. Endemic to western Canary Is. Suggests a dark, slate grey Woodpigeon, with short, rounded wings; lacks white markings. Distinguished from next species by very dark tail with paler grey subterminal band. Adult has vinous breast, reflective pinkish and green neckside patches and orange bill. Juvenile dark brown-grey. Usual call a weak and guttural 4-syllable: *ruor–ruor–ruor–rup* ✦ Laurel forests and tree heath, sometimes in small remnants. Nests in trees, at times on the ground; breeds year-round (peak Nov–May). Consumes woodland fruits. Sometimes visits cultivation ✦ Sedentary, with some wandering ✦ Several thousand birds, latterly increasing; especially numerous in N and E of La Palma and in La Gomera (Garajonay), also common in Tenerife (Anaga massif, La Orotava valley, Los Silos) and relatively scarce in El Hierro. **NT**

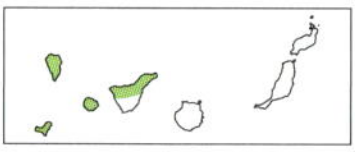

White-tailed Laurel-pigeon *Columba junoniae* (paloma rabiche). ~40 cm. Endemic to western Canary Is. Darker than previous species, with rufous tones and grey tail shading to white distally. Upperparts brownish, nape and neck sides iridescent green; bill whitish. Flies slow, with occasional paused wingbeats. Usual call a trisyllabic: *pu–pu–pooo* ✦ Wooded slopes; in laurel forest, tree heath, mixed pinewoods…, often at lower altitude than previous species. Nests on ground or escarpment, in shaded inaccessible sites; breeds year-round ✦ Sedentary ✦ Uncommon; a few thousand birds, most in C and N of La Palma; common in La Gomera, scarce in Tenerife and very scarce in El Hierro. **EN**

Common Woodpigeon
Dark-tailed Laurel-pigeon
White-tailed Laurel-pigeon

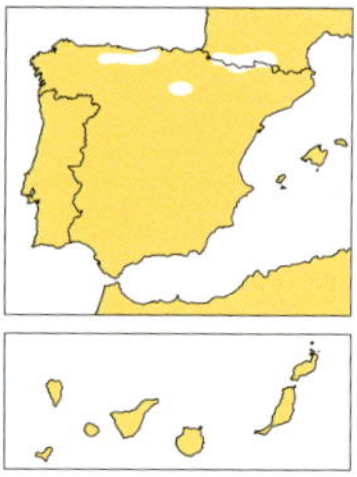

European Turtle-dove *Streptopelia turtur* (tórtola europea). ~27 cm. Summer visitor. Smaller, darker and more contrasted than Collared-dove. Dorsal plumage shows black feather centres with broad orange fringes; greater coverts grey; black and white neckside patches; black and white graduated tail. Juvenile brownish, unmarked. Drowsy coo: *turr–turr, turr–turr…* Often in pairs or small groups ✦ Landscapes with trees, riparian woods, dehesas, orange groves, etc., generally at low altitude. Nests in trees or shrubs. Visits sunflower and cereal crops and drinking places ✦ Summer visitor, arriving fairly late; also passage of European birds(Apr–Jun and Aug–Oct) ✦ Ssp. *turtur*: quite common (hundreds of thousands) but strongly declining (hunted excessively) and now scarce in many areas; spring passage more evident in S and E, autumn passage in W; common in Canary Is. Ssp. *arenicola*: very common in Balearics; breeds in Melilla. **VU**

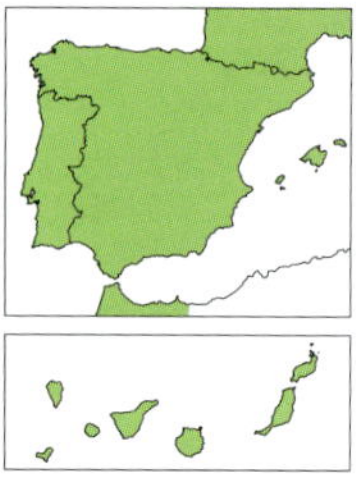

Eurasian Collared-dove *Streptopelia decaocto* (tórtola turca). ~32 cm. Larger and longer-tailed than previous species. Buffish, with some grey and pinkish areas. Thin black half-collar, lacking in juveniles; largely white tail with black base below. Voice, a persistent *cu–cooo–cu* ('de–caoc–to'); also a harsh rrrah. Escaped and feral domestic Barbary Doves smaller, paler and more uniform; lacking pinkish on breast ✦ Chiefly towns, suburbs and parks. Also farms and grain stores ✦ Sedentary, once established ✦ Common (hundreds of thousands) and rapidly increasing; range expanded eastwards across Europe, first reaching Spain in 1960s; initially most numerous in coastal areas. Also in Ceuta and Melilla, Balearics and Canary Is.

SANDGROUSE (Pterocliformes, Pteroclidae) – Terrestrial, medium-sized, granivores, resembling long-winged pigeons with feathered tarsi. Plumages highly cryptic. Nest on ground; chicks precocial. Visit drinking places. Arid and semi-arid regions of Africa, Asia and SW Europe.

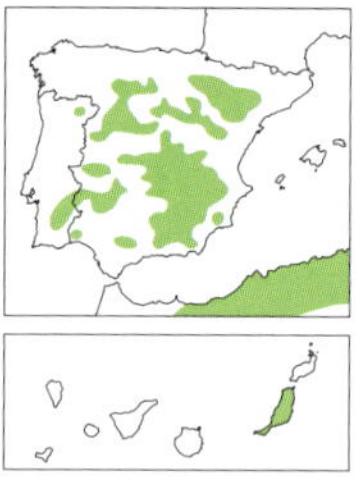

Black-bellied Sandgrouse *Pterocles orientalis* (ganga ortega). 33–39 cm. Call: a rolling *churr…churr–ur–urr*. Bulkier than Pin-tailed, with black belly. Underwings whitish. Males have orange, black-centred throat patch; females barred above, neck and breast spotted. Flight powerful, less whirring than Pin-tailed. Often in pairs or small groups. Visits watering places in summer, ~3 h after dawn ✦ Arid lands: dry croplands with fallows and pastures, set-asides, upland grasslands ✦ Sedentary ✦ Fairly common (thousands, but declining) and widespread; at lower densities than Pin-tailed where the two coincide. Most frequent in Ebro valley, S Meseta and Extremadura. Canary Is: common in Fuerteventura (hundreds) and scarce and irregular in Lanzarote. **VU**

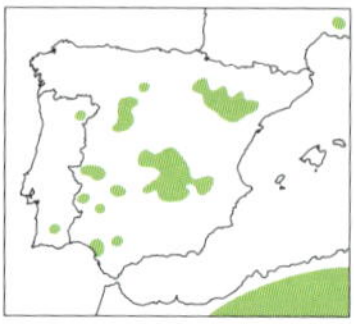

Pin-tailed Sandgrouse *Pterocles alchata* (ganga ibérica). 31–39 cm. Call: *catar…catar…*, or *gaaa…gaaa…* Belly and underwings white; central rectrices elongated. Flight rapid and more pigeon-like than Black-bellied. Males showy at close range; lack dorsal spotting in winter, when throat white and eyestripe indistinct or absent. Females have two neck bars, breast band less rufous, throat always white. Juveniles paler and more uniform, lacking neck bars. Drinks a little earlier than Black-bellied. Often in flocks ✦ Arid plains at low elevations, with cereal crops and areas of dwarf shrub steppe; often on saltflats ✦ Resident, with some nomadism ✦ Uncommon (tens of thousands), declining, and local, most frequent in Aragón and La Mancha. **VU**

European Turtle-dove
Eurasian Collared-dove
♀ ad.
♂ ad.
Black-bellied Sandgrouse
♂ ad.
Pin-tailed Sandgrouse
♀ ad.

NIGHTJARS (Caprimulgiformes, Caprimulgidae) – Crepuscular or nocturnal, long-winged and long-tailed, with highy cryptic plumage. Hunt insects in flight; have small beaks but very large, vibrissae-fringed gapes. Hide during the day. Ground nesters; chicks semi-nidicolous. Variably migratory. Warm and hot regions worldwide.

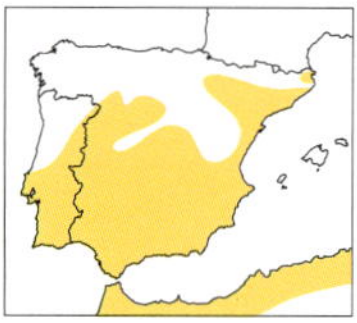

Red-necked Nightjar *Caprimulgus ruficollis* (chotacabras cuellirrojo). ~32 cm. Somewhat larger and longer tailed than European. Warmer-toned, with rufous halfcollar and pale throat and cheek marks. Both sexes have white primary and tail-corner patches. Song: a loud, prolonged, low-pitched series of two-tone notes, recalling a hollow log being struck: *cut-ok–cut-ok–cut-ok...* ✦ Open Mediterranean woodland (holm oaks, pinewoods, junipers), scrub (kermes oak, broom, cistus), orchard groves and set-asides. Generally at low altitudes (<600–1.000 m according to region). Frequently settles on roads and tracks, where often run-over ✦ Summer visitor (Apr/May–Sept/Oct), trans-Saharan migrant ✦ Common (tens of thousands), stable or declining locally. Mediterranean distribution complementary to that of European Nightjar; most abundant in warm dry areas of Extremadura, S Castilla and S Andalucía. Nests in Melilla. Vagrant to Balearic and Canary Is.

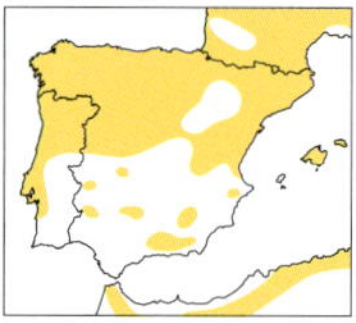

European Nightjar *Caprimulgus europaeus* (chotacabras europeo). ~26 cm. Hard to distinguish from Red-necked with brief views, but somewhat smaller and greyer; only males have white wing and tail patches. Song quite different: a very prolonged churr, wavering in pitch: *RRRR–rrrrr–RRRR–rrrrr...* ✦ Scrubby woodland edges and clearings, or open areas with scattered trees. Especially on mountains, where often above 1,500 m. Frequently settles on roads and tracks, where sometimes run-over ✦ Summer visitor and migrant (Apr–May and Sept–Oct), wintering S of Sahara ✦ Ssp. *meridionalis*: breds at varying densities but generally uncommon (tens of thousands), declining; most abundant in N; in C and S largely limited to high or cold regions. Small Balearic population. Ssp. *europaeus*: migrant. Vagrant to Canary Is.

SWIFTS (Apodiformes, Apodidae) – Highly aerial insectivores, with some similarity to swallows. Wings long, bills short, gapes very wide, all toes short and point forward. Plumages dark. Nests in holes in cliffs and buildings. Many species undertake long migrations. Cosmopolitan.

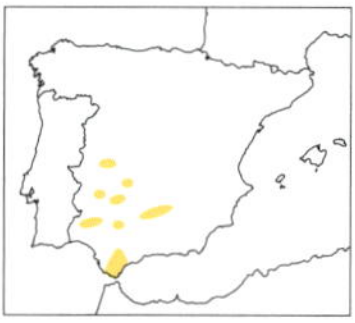

White-rumped Swift *Apus caffer* (vencejo cafre). 14 cm. Very local nester in SW sierras. Slim, black, with narrow white rump patch. Tail deeply forked, often held closed. The white rump of this and the Little Swift may recall House Martin but both swifts are black below ✦ Hilly areas. Takes-over Red-rumped Swallow nests, adorning the entrance with white feathers and down ✦ Summer visitor (May–Oct) ✦ Some tens of pairs nest in S Cádiz and other parts of Andalucía and Extremadura. Recent colonist (first recorded in 1964), may be gradually spreading northwards. Vagrant to Canary Is. **VU**

Little Swift *Apus affinis* (vencejo moro). 12 cm. Very local, in Andalucía. Chunky, black with unforked tail and large, squarish, 'wrap-around', white rump ✦ Nests on buildings, walls and rock ledges, usually in colonies ✦ Partial migrant, some overwinter ✦ Ssp. *galilejensis*: common in Morocco, a few pairs have nested in Cádiz province since 2000; also recorded in the Sierra Morena; vagrant elsewhere (mainly Apr–Jul). Vagrant to Canary Is.

Red-necked Nightjar

European Nightjar

White-rumped Swift

Little Swift

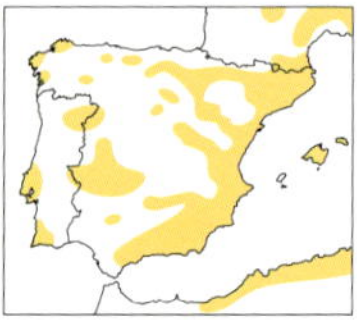

Alpine Swift *Tachymarptis melba* (vencejo real). ~21 cm. Largest swift and the only one with a white belly. Voice is a long, slow but often accelerating trill ✦ Small colonies nest on sea and inland cliffs of diverse substrate and at variable altitudes. Locally on bridges, dams and churches and, increasingly often and especially in Catalonia, on buildings in towns and cities ✦ Summer visitor (Mar/Apr–Sept/Oct), wintering in sub-Saharan Africa ✦ Uncommon (a few tens of thousands), perhaps increasing. Patchily distributed, according to nest cliff availability: commoner in E half, in limestone regions; less so in C (plains) and N. Very local in Mallorca and Menorca. Breeds in Melilla. Regular but scarce migrant in Canary Is.

Common Swift *Apus apus* (vencejo común). ~17 cm. The abundant swift of towns and villages in spring and summer. Has dark belly and scythe-shaped wings, unlike swallows and martins (p. 172) and tends to fly higher. Dark brown, looking black at a distance. Juveniles show pale feather edging, especially above, as in other swifts. Gregarious, frequently forming tight and noisy screaming parties ✦ Nests in holes in buildings and sometimes also on cliffs and in trees (in woodpecker holes in mountain pines; palms). Wide-ranging feeding movements reach mountains, forests, farmland, wetlands… ✦ Summer visitor and migrant (esp. Apr/May–Aug/Sept), trans-Saharan ✦ Abundant everywhere (many hundreds of thousands of pairs) but less so in N. Nests in Ceuta and Melilla, and some nest in Canary Is (at least in Gran Canaria), where also common on passage.

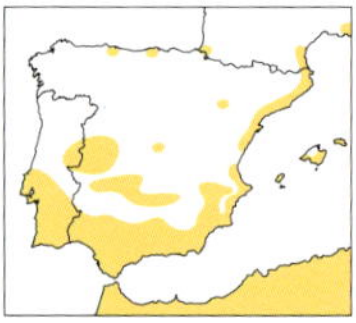

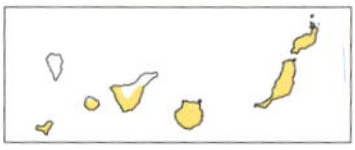

Pallid Swift *Apus pallidus* (vencejo pálido). ~17 cm. Identification needs good light since may otherwise look darker (and Common Swift paler). Shows contrast between paler inner wing and leading edge, outer primaries looking darker. Pale fringes to flank feathers visible in good light. Face white, so eyes stand out; white throat patch larger and wingtip blunter than in Common. Voice somewhat more melodious and disyllabic than Common ✦ Sea cliffs and coastal towns, but also occurs inland (buildings, bridges) where may be overlooked among Common Swifts ✦ Trans-Saharan migrant (Mar–Sept/Oct) occasionally recorded in winter ✦ Ssp. *brehmorum*: common (tens of thousands) on E and S littorals, with some penetration inland (mostly in Andalucía, but also in Extremadura, Castilla-La Mancha, Madrid); few records from N but nests at least in cities of Zaragoza, Santander and Gijón, locally in Galicia. Very common in Ceuta and Melilla. Also nest in Balearics, commonly, and Canary Is, scarce.

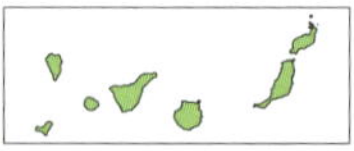

Plain Swift *Apus unicolor* (vencejo unicolor). ~15 cm. The usual swift in the Canary Is (otherwise known for certain to nest only in Madeira). Resembles Common in coloration but smaller, throat dark and tail fork deeper, and more agile and erratic in flight ✦ Nests generally at low altitudes in cliffs, escarpments, villages and towns, but visits all habitats (pinewoods, scrublands…) and then reaches high in mountains ✦ Mainly summer visitor (Jan/Mar–Sept/Oct) but seen occasionally in other months ✦ Some thousands of pairs, perhaps declining. On all islands and main islets, most abundant in Tenerife and Gran Canaria and comparatively rare in Lanzarote and Fuerteventura. **DD**

Alpine Swift
Common Swift
Pallid Swift
Plain Swift

CUCKOOS (Cuculiformes, Cuculidae) – Terrestrial or arboreal, generally medium-sized insectivores. Short-legged with zygodactyl feet (two toes forward, two back). May parasitise nests of other birds. Widespread globally.

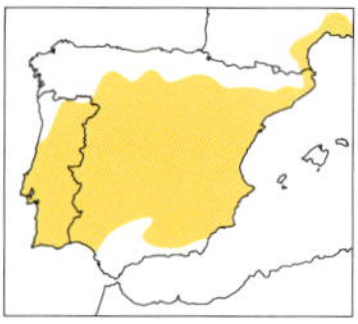

Great Spotted Cuckoo *Clamator glandarius* (críalo europeo). ~38 cm. Larger and longer-tailed than Common Cuckoo, showing white spotting above, on scapulars and wing coverts. Unbarred white below, creamy on throat; cap grey. Juveniles have black caps and vivid chestnut primaries; primary bases still brownish in 2nd years. Noisy; gives a guttural *gah-gah-gah...gak-gak-gak... ko-ko-ko-ko*, falling in pitch and accelerating. Often pursued by Common Magpies ✦ Mediterranean landscapes with some trees (pinewoods, dehesas, almond groves); normally at low altitudes. Parasitises nests of Common Magpies and sometimes other corvids. Eats caterpillars (chiefly of processionary moths) ✦ Summer visitor (Feb/Apr–Jul/Sept), returning early from Africa (midwinter records) ✦ Quite common (tens of thousands), absent from humid regions and variably numerous elsewhere, most common in S and C. Rare migrant in Canary Is; occasional in Balearics.

Common Cuckoo *Cuculus canorus* (cuco común). ~34 cm. More often heard than seen, thanks to the far-carrying familiar song, *cuc–koo*. Wings and tail long, barred below and could be confused with Sparrowhawk (p. 134) but has pointed wings and distinctive flight with wings beating only below the horizontal. Male grey with plain throat. Female may be grey, with barring and rufous colouring on throat or, more rarely, rufous overall. Juveniles grey or variably rufous, with white tips to upperparts and white nape patch ✦ Wide range of wooded or shrubby habitats, up to >2,000 m. Prefers landscapes with copses of oaks and other deciduous trees. Parasitises a diversity of bird species ✦ Summer visitor (Mar/Apr–Aug/Sept, adults departing Jun/Jul), with passage of European birds ✦ Ssp. *bangsi*: common (some hundreds of thousands) and very widespread, most frequent in N and mountains and scarce in hot and barren areas; common in Balearics. Irregular and scarce on passage in the Canary Is. Ssp. *canorus* (from N and C Europe): scarce migrant.

RAILS (Gruiformes, Rallidae) – Small or medium-sized birds, typical of marshland vegetation. Body flattened bilaterally; wings short and flight appears weak; tail short, often moved up and down; toes generally long and thin, sometimes lobed (in coots). Nests on ground within vegetation; chicks nidifugous. Cosmopolitan.

Corncrake *Crex crex* (guión de codornices). ~25 cm. Rare and only a passage migrant, dubbed 'King Quail'. Resembles quails (p. 16) in coloration and skulking habits but larger and takes off more slowly, showing rufous wings and long, dangling legs. Seldom seen except by hunters with dogs. Song chiefly nocturnal, a very loud, rasping, repeated, bisyllabic *crek–crek, crek–crek*, sometimes heard from migrants ✦ Meadows, wet grasslands, alfalfa (most in Europe nest in wet meadows) ✦ Passage (around Mar–Jun and Aug–Nov, peaking in Apr and Sept–Oct) of birds nesting from C France northwards. Some records in other months ✦ Once relatively common but now rare or very rare, following a catastrophic decline in W European populations; perhaps most frequent in E and S of the Peninsula. Vagrant to Balearic and Canary Is. NT

juv.
Great Spotted Cuckoo
♂
Common Cuckoo
♀ grey morph
♀ rufous morph
Corncrake

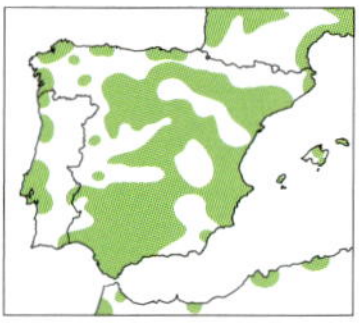

Western Water Rail *Rallus aquaticus* (rascón europeo). 25–28 cm. Largest and commonest of the small rallids. Bill long, red in adults; flanks barred, breast grey. Juveniles pale-headed, with throat white, brownish breast and dull bill. More heard than seen but sometimes appears in creeks and marsh fringes. Vocalisations loud and variable; most often heard at night; some recall squeals of a pig being killed; calls include a harsh, metallic *kik, kik, kik*. Flies weakly, with legs dangling ✦ Inundated marsh vegetation, often at small wetlands, in lakes, ponds, streams, ditches, estuaries, marshes ✦ Sedentary; small numbers of European birds arrive to winter ✦ Fairly common (tens of thousands) and widespread; sometimes reach high densities, especially in N. Avoids mountains and very local in semiarid regions. Nests in Balearics (chiefly Menorca and N Mallorca). Vagrant to Canary Is.

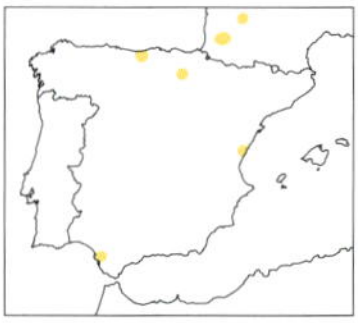

Spotted Crake *Porzana porzana* (polluela pintoja). ~23 cm. Nearly as large as Water Rail but bill short, yellow, with a red mark at base. Plumage spotted, undertail coverts plain, cream. Young lack grey on head and have fine breast spotting. Skulking, solitary and hard to surprise, as other crakes. Nocturnal song distinctive: the far-carrying 'whiplash' call, an ascending *whitt*, repeated every second for long periods; migrants may sing ✦ Marsh vegetation in marshes, lagoons, ricefields ✦ Migrant (from Mar–Apr and Aug–Oct), with rare winter and breeding season records ✦ A few pairs seem to nest, very locally. Commonest crake on passage, relatively frequent on Mediterranean littoral and more so in spring than autumn. Scarce but regular in Canary Is. **DD**

Little Crake *Zapornia parva* (polluela bastarda). 18–20 cm. Spanish name refers to resemblance to Baillon's Crake in size and barred undertail coverts and to Spotted Crake in having green legs and red bill base. No lateral barring; brownish olive above, largely lacking white; primary projection notable. Only crake where female plumage differs from male, being buffy-grey below. Juvenile recalls female but paler, lacks red on bill. Male sings mainly at night, a loud, nasal *quek–quek–quek*, repeated for up to several minutes before accelerating and descending in pitch, ending on softer, more guttural notes ✦ Marsh vegetation ✦ Passage migrant (Feb–May and Aug–Nov) ✦ The rarest crake in Spain: hardly any definite breeding records and uncommon on migration (very few breed in W Europe). Most records are in spring in the NE quadrant, chiefly Catalonia. Rare and irregular in Canary Is. **DD**

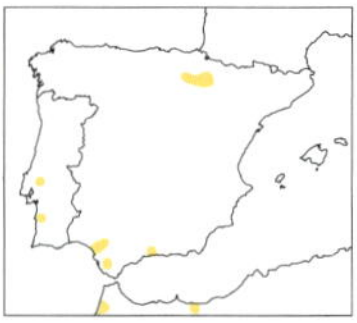

Baillon's Crake *Zapornia pusilla* (polluela chica). ~18 cm. Small, with barred undertail coverts. Bill green, lacks red. Plumage resembles Water Rail; face and breast grey, flanks densely barred. Rufous-brown above, with irregular white blotching; primary projection much shorter than in Little Crake; legs variably pinkish or greenish. Juvenile has barred breast and yellowish bill. Very skulking. Peculiar song may be confused with a croaking frog: a series of rasping rattle, lasting 2–3 seconds and repeated every 1–2 seconds ✦ Littoral marshes, ricefields, lakes, ponds ✦ Summer visitor (Feb/Apr–Jul/Oct) with some rare winter records ✦ Ssp. *intermedia*: breeding population small (probably only tens of pairs), apparently declining and fluctuates according to water levels; chiefly in S and E (Marismas, Delta), rare inland and sporadic in N. Negligible passage (species virtually absent from W Europe). Vagrant to Balearic and Canary Is. **DD**

Western Water Rail
Spotted Crake
♀
♂
Little Crake
Baillon's Crake

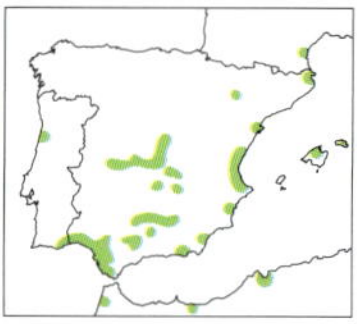

Purple Swamphen *Porphyrio porphyrio* (calamón común). 45–50 cm. Like a huge Moorhen; adults purple and indigo; juveniles grey-blue; bill, legs and frontal shield red. Bill very large; legs large and long-toed. Tail often cocked to show conspicuous white undertail coverts. Various vocalisations including frequent nasal trumpeting ✦ Marginal vegetation (reeds, reedmace) in marismas, marshes, lakes, reservoirs, ricefields and sluggish rivers ✦ Sedentary, with dispersive and drought-avoidance movements ✦ Locally common and spreading since the 1980s, aided by reintroductions (Aiguamolls and albuferas de Valencia and Mallorca), after surviving a critical period in the 1960s. From the Marismas (some thousands of pairs) has recolonised the upper Guadalquivir and lagoons and reservoirs in La Mancha, Madrid and Extremadura. Common at the Delta, now nests at Sariñena (Aragón) and in Menorca.

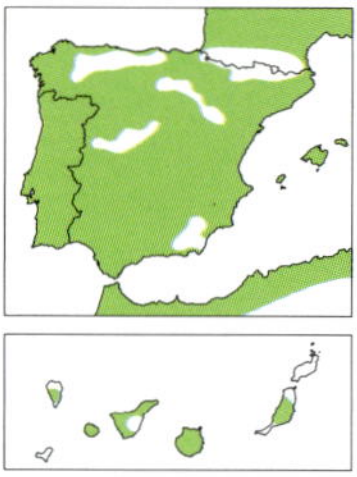

Common Moorhen *Gallinula chloropus* (gallineta común). ~32 cm. The familiar 'waterhen'. Looks black at a distance, with white flank markings, exhibiting black and white stern by frequent tail jerking; bill and frontal shield red; legs green with red 'garters'. Juvenile greyish with dull beak. Chicken-like on land, swims buoyantly with jerking head; flies well though infrequently. Characteristic calls include a loud, explosive, *kurrrruk*, *kur-rik* or *kark* ✦ Wetlands, large or small, up to ~1,000 m up (rivers, streams, ditches, pools, lakes, gravelpits, parks) ✦ Resident, with some passage and wintering by European birds (Mar/Apr–Sept/Nov) ✦ Very common (tens of thousands of pairs), increasing. Most abundant in rice-growing areas, scarce in semiarid and montane regions. Spreading in Canary Is (commonest in Tenerife, Gran Canaria and Fuerteventura).

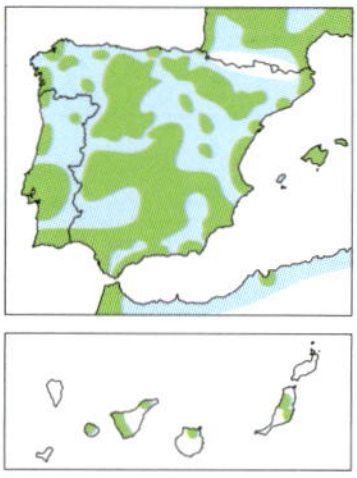

Common Coot *Fulica atra* (focha común). ~38 cm. Common. A black waterbird with a white bill, only confusable with Red-knobbed Coot. Juvenile greyer, lacking bill shield; face and neck whitish as in young Moorhen but lacks white stern and flank marks. Secondaries show whitish edge in flight. Often gives a monosyllabic *kowk* or explosive *piit piit* calls ✦ Freshwater wetlands (lakes, reservoirs, marshes, marismas, slow rivers) generally below 1,000 m. Nests in fringing vegetation and eats submerged plants ✦ Spanish birds essentially sedentary, with dispersive and drought-avoidance movements. Large numbers of birds from C and N Europe occur on passage and in winter (Oct–Mar) ✦ Widespread and common breeder (tens of thousands), perhaps increasing although numbers fluctuate, with notable concentrations at some major wetlands (Marismas, Delta, La Mancha). Abundant in winter (some 100,000 censused). Has recently colonised Canary Is (<50 pp.).

Red-knobbed Coot *Fulica cristata* (focha moruna). ~40 cm. Rare. Resenbles Common Coot, with red 'knobs' only evident when breeding. Bill bluish (looks 'dirty' at a distance) contrasts with white frontal shield; lacks feathered wedge at bill base of Common Coot and longer-necked, squarer-headed and shows squarer stern when swimming; secondaries lack white edge. Juveniles only paler on throat. Calls distinct, including a sharp, shrill *krik*, a shrill, trilled *krrt* and a double clucking note ✦ Wetlands with abundant submerged vegetation ✦ Sedentary, with dispersive and drought-avoidance movements (probably to Morocco) ✦ Formerly commoner but now very rare (tens) and restricted to Marismas and a few Andalusian lagoons; reintroductions attempted in Valencia, Catalonia and Balearics. **CR**

juv.
Purple Swamphen
Common Moorhen
juv.
juv.
Common Coot
1st winter
ad. winter
1st winter
ad. winter
Red-knobbed Coot

CRANES (Gruiformes, Gruidae) – Long-legged, long-winged birds native to Eurasia, Africa, Australia and North America. Bushy elongated tertials comprise a false tail.

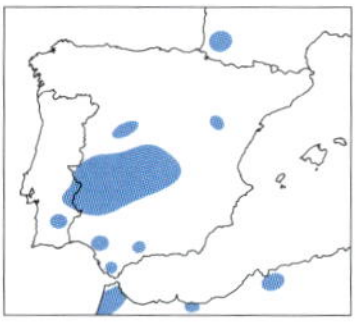

Common Crane *Grus grus* (grulla común). 110–120 cm. Very large and grey (as Grey Heron, p. 68, and Greylag Goose, p. 20, flies with neck extended, in V-formations, calling frequently with a deep, rolling *krroo*. Juveniles brown-headed. Family groups stay together in flocks. Has a measured, stately gait; majestic in flight, alternating wingbeats and glides; soars. Migrates chiefly by day ✦ Holm oak dehesas, young crops, stubbles ✦ Scandinavian and Baltic birds occur in winter and on passage (Oct/Dec to Feb/Mar) ✦ No longer nests. Common in winter (many tens of thousands), steadily increasing; chiefly in Extremadura (esp. in Vegas Altas del Guadiana) and neighbouring regions; crosses peninsula diagonally to and from the Pyrenees and France, the main staging point being the laguna de Gallocanta. Scarce in Balearics. Vagrant to Canary Is.

BUSTARDS (Otidiformes, Otididae) – Large terrestrial birds of Africa, Eurasia and Australia. Toes short, hindclaw absent. Cryptic. Gregarious. Omnivorous. Courtship displays spectacular. Nests on the ground; chicks nidifugous.

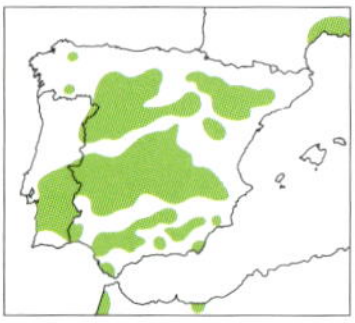

Little Bustard *Tetrax tetrax* (sisón común). 40–45 cm. Presence revealed in Apr–May, at sunrise and sunset, by the male display call, a frequent, far-ranging, snorted *prrit*. Males larger, with distinctive neck markings only in breeding plumage. Flight silhouette ducklike; wings, showing much white, in males make a whistling sound. On ground resemble large, long-necked partridges ✦ Extensive croplands and pastures (with fallows, set-asides and legumes); also on irrigated crops in winter (alfalfa) ✦ Resident and dispersive; some (few) arrive from France to winter ✦ Common (tens of thousands) but in rapid and worrying decline. Common in S Meseta and in Extremadura; uncommon or scarce in the Duero, Ebro and Guadalquivir valleys; rare elsewhere. Vagrant to Balearic and Canary Is. **VU**

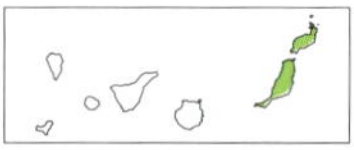

Houbara Bustard *Chlamydotis undulata* (avutarda hubara). 55–65 cm. Canary Is. Medium-sized, graceful, long-tailed; runs fast. Crown pale. Ornamental plumes, longer in males and lacking in juveniles, form black line down neck. Long, narrow wings show large white primary patch; wingbeats slow ✦ Semiarid plains, more rarely on lava fields; visits crops (legumes, alfalfa) ✦ Resident but wanders ✦ Ssp. *fuertaventurae*: common (many hundreds distributed between Fuerteventura and Lanzarote, with some on La Graciosa islet). **EN**

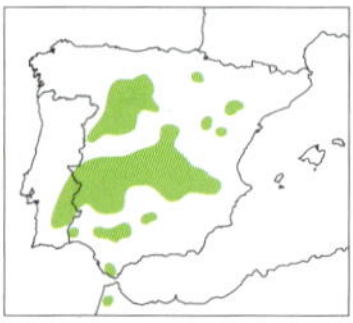

Great Bustard *Otis tarda* (avutarda común). 75–105 cm. Males enormous. In heavy, deliberate flight, shows much white on broad wings and bulky breast. Breeding season males show thick bluish necks, long 'beards' and raised tails; spectacular displays at traditional leks in Mar–Apr convert them into gigantic 'snowballs'. Juveniles have browner heads and necks. Form small flocks ✦ Open plains with extensive non-irrigated crops (cereals and leguminous crops) ✦ Resident, but dispersive or nomadic ✦ Quite common (>30,000 birds) and increasing, but tending towards range contraction. Chiefly on the N Meseta (notably Villafáfila and Tierra de Campos) and Extremadura (Llanos de Cáceres), La Mancha and Madrid. Very few now elsewhere. **VU**

juv.
Common Crane
♀
Little Bustard
♂ breeding
Houbara Bustard
♀
♂
Great Bustard

DIVERS (loons) (Gaviiformes, Gaviidae) – Large, fish-eating, foot-propelled diving birds. Bill dagger-shaped. Legs lobed and set well back. In flight show drooping neck and projecting feet. Recall grebes but thicker-necked, have rounded sterns and lack white on wings. Allied to petrels and penguins. Nest at high latitudes of Eurasia and North America. Scarce winter visitors to Spain, chiefly to northern coastal waters. Usually seen in drab winter garb but late spring birds may show moult into breeding plumage.

Red-throated Diver *Gavia stellata* (colimbo chico). 55–65 cm. Smallest and dullest diver, with dorsal spotting visible only at close range. Extensive white on neck and face, extending above eyes which stand out, although juveniles retain fine streaking on the face and necksides until midwinter. Distinctive head profile, with flat crown almost level with uptilted bill. Flat-chested ✦ Shallow coastal waters, especially bays and estuaries, sometimes off beaches ✦ Winter visitor (Nov–Mar, most Dec–Jan). Nearest breeders in Scotland and Scandinavia ✦ Perhaps rarest diver in Spain (a few dozen). Regular on Biscay coast (Bidasoa and Nervión estuaries; Santoña, Santander and Gijón bays) with a very few in W Andalucía and Catalonia; irregular or occasional elsewhere. Vagrant to Balearic and Canary Is.

Black-throated Diver *Gavia arctica* (colimbo ártico). 60–70 cm. Intermediate-sized, although overlaps with Red-throated. Shapely and elegant, curve of head continuing down neck. Straight, fairly stout bill, horizontal when swimming. Two-tone in winter, with clearly demarcation of dark and light on head and neck (often bordered darker) and pale flank patch; eye does not stand out. Full-breasted. Juvenile with pale feather edging above ✦ Sheltered coastal waters (bays, inlets) ✦ Winter visitor (Nov–Apr, peaking Dec–Jan). Nearest breeders in Scotland, Norway and Sweden ✦ Regular but scarce (a couple of hundred) in Galicia (most), the Biscay coast and N Catalonia (Roses bay); very infrequent elsewhere in the Peninsula. Vagrant to Balearic and Canary Is.

Great Northern Diver *Gavia immer* (colimbo grande). 70–90 cm. Largest and heaviest diver. Large head emphasises steep forehead that may be raised to form a 'bump'. Pied appearance in winter, as Black-throated, but black and white areas less clearly demarcated and black forms a dark half-collar on the lower neck; eye more evident. Robust bill, grey although may appear whitish, but culmen and cutting edges always dark. Juvenile with pale scaly pattern above, often well into winter ✦ Estuaries, bays, inlets, harbours; sometimes exposed rocky shores ✦ Winter visitor (Nov–Apr, most Dec–Jan) originating from N America, Greenland and Iceland ✦ Overall the least scarce diver in Spain (a few hundred), probably declining, and concentrates chiefly in Galicia and the Biscay coast (esp. Asturias) and rare or exceptional in Mediterranean; occasionally observed on reservoirs inland. Vagrant to Balearic and Canary Is. **VU**

juv.
Red-throated Diver
breeding
ad. winter
breeding
Black-throated Diver
ad. winter
breeding
Great Northern Diver
ad. winter

STORM-PETRELS (Procellariiformes, families Oceanitidae –*Oceanites* and *Pelagodroma*– and Hydrobatidae –*Hydrobates*) – Very small, short-billed, pelagic petrels. Most dark with white rumps. Plankton eaters. Nest on islets, attending at night. Lay one egg. Their tideline corpses may appear on beaches after storms. Tricky to identify.

Wilson's Storm-petrel *Oceanites oceanicus* (paíño de Wilson). ~17 cm. Seldom reported. Intermediate-sized. Legs long and project beyond tail in flight; yellow webs hard to see. Wings somewhat rounded, with a pale covert panel above and dark below. Broad white rump. Flight direct, wingbeats interspersed with sometimes long glides; swallow-like. Follows ships. Feeds with legs dangling and wings in shallow V ✦ Usually far from land ✦ Nests in Antarctica and round the Southern Ocean and moves as far as the N Atlantic during the austral winter (May–Sept) ✦ Rarely seen near coasts but probably regular inshore (mainly in Galicia Aug–Sept). Occasional in the Canary Is.

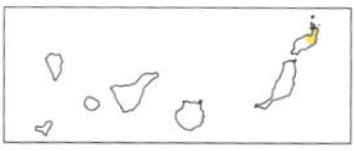

White-faced Storm-petrel *Pelagodroma marina* (paíño pechialbo). ~20 cm. Canary Is. Distinctive, with white face and underparts. Legs long, often dangled. Wings broad-centred; butterfly-like. Bouncy flight distinctive ✦ Burrows in sandy ground ✦ Nests mainly Mar–Jul, apparently absent Oct–Dec. Large nearby colonies in Selvagens and Cape Verde Is ✦ Ssp. *hypoleuca*: a few dozen pairs in Montaña Clara, Alegranza and perhaps other islets of Lanzarote. **VU**

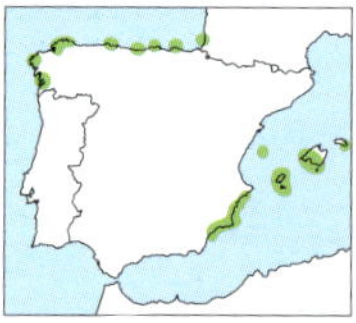

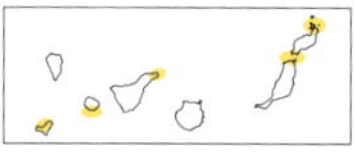

European Storm-petrel *Hydrobates pelagicus* (paíño europeo). ~15 cm. Smallest and sole breeding species in mainland Spain and Balearics. Square-tailed; large white rump. Wings short, nearly dark above but with a distinctive broad white band below. Flight fluttering; bat-like, with short glides. Foot paddles and raises wings when feeding. Follows ships. Male has a purring song interspersed with grunts, given from nesthole ✦ Nests in burrows, crevices or caves, on rat-free islets ✦ Chiefly summer visitor; some seen in winter. Nests Apr–Oct ✦ Fairly common (a few thousand). Ssp. *pelagicus*: Biscay coast, Galicia (scarcer in W) and Canary Is (most on Alegranza). Ssp. *melitensis*: Mediterranean, off mainland (most on Benidorm islet) and Balearics (most in Ibiza). **VU**

Madeiran Storm-petrel *Hydrobates castro* (paíño de Madeira). ~20 cm. Canary Is. Similar to Leach's but white rump much more evident, paler upperwing panel less distinct and tail not obviously forked; looks square. Flight more direct ✦ Rocky islets ✦ Nests in burrows, crevices or caves ✦ Year-round; nests in winter ✦ Some hundreds of pairs nest on the Anaga rocks, Alegranza islet and elsewhere. Vagrant to mainland Spain (p. 227) but some hundreds of pairs nest on Berlengas Is, just off W mainland Portugal. **EN**

Leach's Storm-petrel *Hydrobates leucorhous* (paíño boreal). ~20 cm. Larger than European; longish, flexed, pointed wings (recall a marsh tern). Tail clearly forked. White rump comparatively inconspicous; may appeared V-shaped, with an indistinct darker central line. Pale upper wingband conspicuous, contrasting with black remiges. Distinctive bounding flight, with frequent changes of direction. Feeds on horizontal, arched wings, without foot paddling ✦ Pelagic ✦ Summer and autumn. Nests in British Isles, Iceland and N America ✦ Frequent inshore after Atlantic storms. Regular in Canary Is. Occasional in the Mediterranean.

Wilson's Storm-petrel
White-faced Storm-petrel
European Storm-petrel
Madeiran Storm-petrel
Leach's Storm-petrel

PETRELS AND SHEARWATERS (Procellariiformes, Procellariidae) – Pelagic or coastal, only coming to land to nest. Fly on stiff wings, with rapid, mechanical wingbeats alternating with sea-skimming glides. Eat fish and cephalopods. Follow fishing boats. Nest in burrows and cavities between rocks, visiting nests at night. Lay one egg.

Northern Fulmar *Fulmarus glacialis* (fulmar boreal). 45–50 cm. Recalls a gull, with petrel-style flight. Body and neck stout; bill short; dark eye stands out; pale primary patch ✦ Seldom seen from land. Tideline corpses may appear after winter storms ✦ Foraging or wintering birds recorded year-round. Nearest breeders in N France and British Isles ✦ Sometimes recorded at sea in the Bay of Biscay (esp. Jul–Sept), rare further south. Vagrant to Canary Is.

Great Shearwater *Ardenna gravis* (pardela capirotada). 43–51 cm. Summer/autumn. As large as some Cory's, but slimmer, with black cap and white collar. Shows brown collar patches and a 'dirty' brown belly patch; some dark barring and spotting on inner underwing. Bill black. Flies on stiff wings, with rapid wingbeats ✦ Less coastal than other shearwaters ✦ Nests in S Atlantic (Tristan da Cunha) and skirts W Europe in Jul–Nov after visiting N America ✦ Common at sea off Galicia and Asturias, but seldom seen from land; most often inshore with strong NW winds (Cape Bares, esp. Sept–Oct). Scarce in Cantabria and Basque Country; rare or occasional off W Andalucía. Vagrant to the Mediterranean. Rare but regular in Canary Is.

Sooty Shearwater *Ardenna grisea* (pardela sombría). 45–50 cm. Summer/autumn. Very dark overall (but beware poor light sightings of other shearwaters, esp. Balearic, or Arctic Skua, p. 112); underwings silvery. Medium-sized. Body plump and tapering, wings narrow, often backswept at wrists. 'Towers' high suddenly with strong winds ✦ Quite often seen from coasts ✦ Breeds in sub-Antarctic waters and skirts W Europe in Jul–Nov ✦ Numerous off Galicia and Asturias (Cape Bares, Cape Peñas) and common elsewhere off Biscay coasts (esp. Sept–Oct). Scarce off W Andalucía; occasional in Mediterranean and Canary Is.

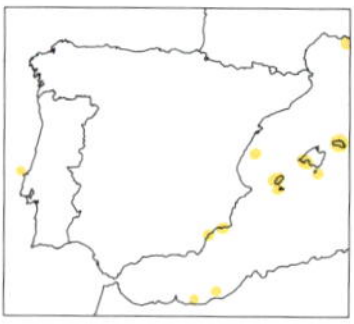

Scopoli's Shearwater *Calonectris diomedea* (pardela cenicienta mediterránea). 44–49 cm. Robust, brownish above and white below, with conspicuous yellowish bill. Darkish head and neck give hooded appearance (compare Great Shearwater). Back mottled paler. Uppertail coverts variably white. Flight more deliberate and relaxed than other shearwaters, with deliberate flexible wingbeats in calm weather. Forms large rafts at dusk off breeding sites, visiting nests after dark when calls noisily. Call bisyllabic: *kaaa–wah, kaaa–wah* ✦ Nests on uninhabited islands and islets. Fishes in coastal waters ✦ Summer visitor (Feb/Mar–Oct/Nov), with some winter records ✦ Common in Mediterranean (thousands of pairs), most in Balearics but also on Chafarinas, Columbretes and Murcia islets; declining; **EN**. Passage through Strait very noticeable at Ceuta (hundreds of thousands). Regular on passage in Canary Is.

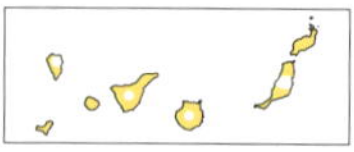

Cory's Shearwater *Calonectris borealis* (pardela cenicienta canaria). 48–56 cm. Treated as ssp. of *C. diomedea* until recently. Larger and darker-headed, with more extensive dark underside to primaries. Call trisyllabic (bisyllabic in *diomedea*). Also very similar to **Cape Verde Shearwater** *C. edwardsii* (pardela cenicienta caboverdiana); smaller, with darker, more slender bill; occasional in Canary Is ✦ Nests on islets and on cliffs and escarpments of the main islands ✦ Summer visitor (Feb–Oct) ✦ Most are on the Azores, Madeira and Canary Is but there is a large colony on the Berlengas, W. Portugal and smaller ones in Galicia; a few nest in the Mediterranean on the Columbretes, Chafarinas and islets off Almería. Very common off Galicia and Biscay coasts, and in the Strait, in summer/autumn. Abundant in Canary Is (tens of thousands), declining; **VU**

Northern Fulmar
Great Shearwater
Sooty
Shearwater
Scopoli's Shearwater
Cory's Shearwater

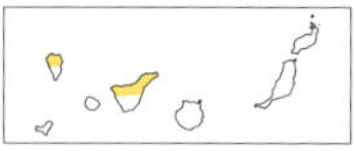

Manx Shearwater *Puffinus puffinus* (pardela pichoneta). 30–35 cm. Atlantic only. Medium-sized. Near-black upperparts and strikingly white underparts, including undertail coverts, sharply defined. Small, white, post-auricular, half-moon patch. Head small; bill long, slender. Flight agile. Often in small flocks ✦ In Canary Is nests on often inaccessible escarpments in laurel forests ✦ Summer visitor to N Atlantic (most nest in British Isles), with postnuptial passage in Aug–Nov. Occasional in winter ✦ Nests in Canary Is (tens of pairs., at least in N and NE of La Palma and N Tenerife). Abundant off Galicia and Biscay coast in summer/autumn, but less so than Balearic Shearwater. Very scarce in Gulf of Cádiz and vagrant in the Mediterranean. **EN**

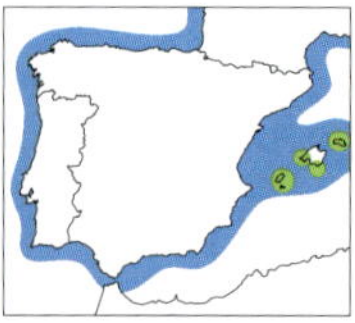

Balearic Shearwater *Puffinus mauretanicus* (pardela balear). 34–38 cm. Nests only on Balearics. Larger than Manx and Mediterranean Shearwaters. Dark brown above and variably dirty white below, poorly contrasted with diffuse neck shading; cloacal area and undertail coverts dark. Highly gregarious ✦ Approaches coasts. Nests on islets, and on cliffs of main islands (Formentera, Mallorca), mainly in caves ✦ Year-round. Nests Mar–Jun; with moult migration to Atlantic France (May–Oct), although many remain in Mediterranean; returns Sept–Jan ✦ A few thousand pairs (mainly in Formentera), declining. Concentrates off Catalonia/Valencia (Dec–Jan). Strong passage through Strait (Jun). Present year-round in Atlantic but most frequent in summer/autumn (Sept–Oct). Vagrant to Canary Is. **CR**

Mediterranean Shearwater *Puffinus yelkouan* (pardela mediterránea). ~35 cm. Mediterranean endemic. Resembles Balearic Shearwater but looks smaller and more compact; shows clearer contrast between dark upperparts and white underparts, especially on neck and face. Usually shows diagonal dark axillary bar and finer line across wing base ✦ Only Spanish colony is on sea cliffs ✦ Present year-round, but seldom reported. Nests mainly Apr–Jul dispersing E (Black Sea) ✦ Recently found nesting in Menorca (Mola de Maó); nearby colonies in Provence and Sardinia. Regular off Catalonia and more sporadically elsewhere in Mediterranean; perhaps reaches Galicia and Biscay coast occasionally.

Audubon's Shearwater *Puffinus lherminieri* (pardela de Audubon). 25–30 cm. Canary Is, uncommon. Small and compact, delicate. Upper and under sides contrast strongly. Dark eye stands out in white face. Bill short, slender and dark. Wings comparatively short, arched. Flight distinctive; wingbeats rapid, almost auk-like; glides little ✦ Nests colonially on islets and inaccessible coastal cliffs ✦ Year-round, most frequent in spring (lays Jan–Mar) and summer. Postnuptial dispersal ✦ Ssp. *baroli* (formerly included in **Little Shearwater** *P. assimilis*, pardela chica): Canary Is, hundreds, at least; La Gomera, Anaga rocks (Tenerife) and Lanzarote islets. Present off Atlantic mainland coasts summer/autumn but very rarely seen from land (p. 228). **EN**

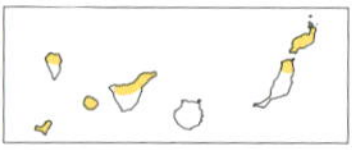

Bulwer's Petrel *Bulweria bulwerii* (petrel de Bulwer). 26–28 cm. Canary Is. Suggests a cross between a shearwater and a storm-petrel, with head like the latter and a long, slender tail. Very dark, with diffuse pale carpal bar. Wings very long, narrow, often arched. Flight agile, distinct from storm-petrels. Calls barking ✦ Nests colonially on islets, in crevices, caves or under rocks ✦ Summer visitor (Feb–Oct; nests May–Sept) ✦ Some hundreds, most on islets N of Lanzarote and on Anaga and Garachico rocks, Tenerife). Vagrant to the Peninsula (p. 228). **EN**

Manx Shearwater
pale
Balearic Shearwater
intermediate
juv.
dark
Mediterranean Shearwater
pale
dark
Audubon's Shearwater
Bulwer's Petrel

STORKS (Ciconiiformes, Ciconiidae) – Very large, long-legged birds. Long-billed and broad-winged, with emarginated primaries. Soar. Diet carnivorous. Worldwide.

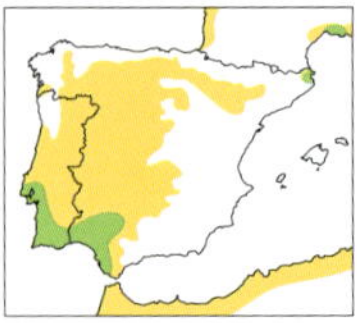

White Stork *Ciconia ciconia* (cigüeña blanca). 100–110 cm. Familiar and unmistakable. Has bill clattering display. Often in flocks ✦ As often at wetlands (marismas, ricefields, riverbanks, lakes, pools) as on meadows, pastures, crops or in dehesas. Frequents rubbish dumps. Nests large; isolated or in loose colonies; on buildings, pylons, trees, etc. ✦ Partial migrant (most Jan–Aug; many year-round); passage of W European birds ✦ Common (tens of thousands), especially in Extremadura and W Castilla y León; Recent increase, following large decline. Widespread but avoids forests and mountains. Many winter, especially in S (Marismas). Large concentrations occur at the Strait, especially during the southward movement (mainly Jul–Aug). Scarce migrants occur in Balearics and Canary Is.

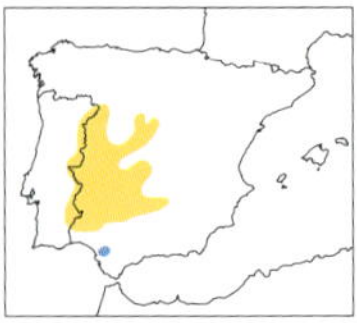

Black Stork *Ciconia nigra* (cigüeña negra). 95–100 cm. Smaller, more elegant and much rarer than White Stork. Coloration distinctive, in flight or at rest. Juvenile lacks gloss; bill and legs greenish. Generally solitary and avoids habitation ✦ Well-wooded hilly country, where fishes in small rivers, streams and pools. Nests on cliffs, often near water, or in forks of large trees (oaks, pines) ✦ Mainly summer visitor (Feb/Mar–Sept/Oct) ✦ Breeding population small (a few hundred) but increasing, in SW and C (Duero, Tagus and Guadiana basins), gradually spreading N and E. Some winter (Marismas). Hundreds of birds, increasing, cross the Peninsula diagonally between the Pyrenees and the Strait, staging at reservoirs. Rare migrant in Balearics. Vagrant to Canary Is. **VU**

IBISES AND SPOONBILLS (Pelecaniformes, Threskiornithidae) – Medium-sized or large wading birds, with curved or spatulate bills. Sociable. Eat invertebrates. Nest colonially, often with herons, in trees or marsh vegetation. Cosmopolitan.

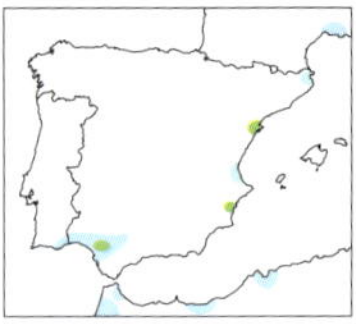

Glossy Ibis *Plegadis falcinellus* (morito común). 55–65 cm. Very dark, looks black at a distance, with decurved bill. Juveniles browner, lack gloss. Flight silhouette distinctive; elongated and slightly hunched; narrow wings look relatively short. Flies in irregular lines or Vs. Consider the **Northern Bald Ibis** *Geronticus eremita* (ibis eremita), recently introduced in Cádiz province ✦ Marshes, deltas, marismas, ricefields ✦ Year-round. Dispersing or migrant birds (from E Europe?) most frequent in Sept–Dec ✦ Uncommon but increasing. Most frequent in E and S but records elsewhere ever more widespread. Formerly extinct as a breeder but has recolonised and there are sizable colonies at Doñana (thousands of pairs), the Delta, Santa Pola, Daimiel and elsewhere. Rare in Balearics and Canary Is. **VU**

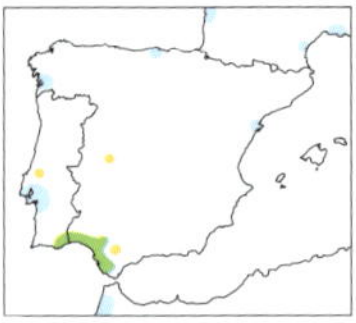

Eurasian Spoonbill *Platalea leucorodia* (espátula común). 80–90 cm. Large and white; recalls egrets (p. 66) but has spatulate bill and flies with neck extended. Breeding adult has yellowish crest and breast patch. Juvenile and 1st winter has black primary tips. Flies in lines. Feeds in fairly deep waters, sweeping bill from side to side ✦ Coastal wetlands and sometimes at lakes or pools inland ✦ Mainly summer visitor, with passage of French and Dutch breeders (~Feb–Apr and Sept–Oct), but some numbers winter ✦ Nests in SW (thousands, fluctuating according to water levels), mainly at Marismas del Guadalquivir and Odiel; now also nests in Extremadura and Ávila. Migrants stage on N coast (Santoña, Guernica, Galician rías) and across W Spain. Rare and irregular in Balearics. Scarce but regular in Canary Is. **VU**

White Stork
ad.
ad.
Black Stork
juv.
winter
Glossy Ibis
ad. summer
Northern
Bald Ibis
ad.
Eurasian Spoonbill

HERONS (Ciconiiformes, Ardeidae) – Slim wading birds with long necks and dagger-shaped bills, adapted for fishing. Sizes vary. Fly with neck retracted and legs extended. Develop ornamental plumes in breeding season, when colours of bare parts may also change. Colonial nesters, often in mixed colonies. Voices harsh. Cosmopolitan, most evident in warm and tropical regions.

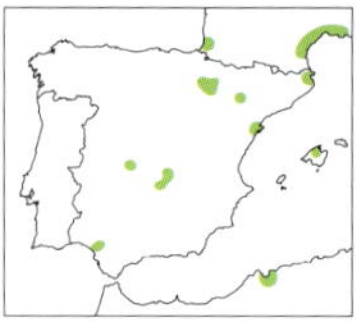

Eurasian Bittern *Botaurus stellaris* (avetoro común). 70–80 cm. Sturdy, thick-necked and relatively short-legged, with highly cryptic plumage. Cap and malar stripe black. Skulks in reedbeds and occasionally seen flying low over them, on broad wings; recalling a large owl. Most easily detected by territorial song of male, often nocturnal; a far-carrying deep, thumping boom, repeated at short intervals ✦ Extensive reedbeds, in freshwater or weakly saline littoral lagoons, lakes, marismas and reservoirs ✦ Partial migrant, may be found year-round; some dispersing European birds occur on passage (Mar–Apr and Sept–Nov) and in winter ✦ Very scarce (a few tens of booming males) and local (chiefly in Ebro valley in Navarra/Aragón). Scarce but more widespread in winter and on dispersal. Rare migrant in Balearics (nests at La Albufera, Mallorca); occasional in Canary Is. **CR**

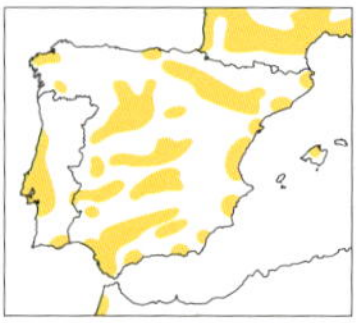

Common Little Bittern *Ixobrychus minutus* (avetorillo común). 33–38 cm. Pigeon-sized with striking bicoloured wings. Skulking, but often seen flying between clumps of marsh vegetation, with rapid wingbeats alternating with glides. Male black-backed; female duller; back brown, streaked black; juvenile more rufous, neck streaked, back pale. Creeps within marsh vegetation; climbs reeds. Voice froglike; male song mainly nocturnal, a muffled barking *grro*, repeated every 2s. Solitary or in pairs. Activity largely crepuscular or nocturnal ✦ Dense vegetation over water (reedbeds, reedmace, sedge tussocks, canebrakes, willows) around lakes and pools, in marshes and along canals and slow-flowing rivers, at low elevations. Nests solitary or in loose colonies ✦ Summer visitor and passage migrant (Apr–Aug/Sept); rare in winter ✦ Local and rather uncommon (a few thousand pairs. Most frequent at coastal sites in Valencia and Catalonia (Albufera, Delta) and the Marismas; scarce but widespread in the main river basins. Nests at the Albufera in Mallorca (tens of pairs). Rare but regular on passage in the Canary Is, has recently nested in Tenerife.

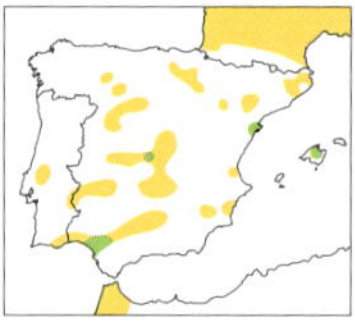

Black-crowned Night-heron *Nycticorax nycticorax* (martinete común). 58–65 cm. Small and stocky, bill stout. Coloration distinctive; grey, face white, cap and back black, legs yellow. Juvenile brownish and less distinctive; recalls Common Bittern but smaller, lack head markings and wings and back are white-spotted. 2nd years have transitional plumage. Roosts in trees by day, emerging to fish at dusk. Wings broad and rounded. Gives a crow-like croak ✦ Chiefly riparian, roosting and nesting in trees (poplars, willows, alders, tamarisks); also in extensive reedbeds. Feeds in rivers, streams, pools, canals, marismas, ricefields… Nests colonially, often alongside egrets ✦ Partial migrant; most are trans-Saharan summer visitors (Mar/Apr–Aug/Oct); hundreds remain in winter and some European birds occur on passage ✦ Uncommon (a few thousand pairs), increasing; numbers fluctuate widely. Nests principally along main river basins (Duero, Tagus, Ebro, Guadalquivir). Winters chiefly in S (Marismas), increasingly. Passage or dispersing birds widespread on mainland and in Balearics (where nests in La Albufera, Mallorca). Regular but scarce in Canary Is.

Eurasian Bittern
♂ ad.
Common Little Bittern
♀ ad.
ad.
juv.
Black-crowned Night-heron

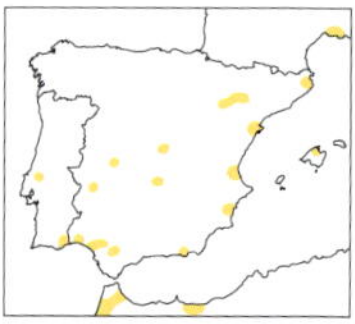

Squacco Heron *Ardeola ralloides* (garcilla cangrejera). 40–48 cm. Somewhat smaller than Cattle Egret, buffish overall, with long, black-tipped bill. Mantle and scapulars contrast with white wings and tail in flight; legs protrudes lightly; wingbeats quite rapid. Breeding adults have bill base blue and long ornamental nape plumes. Juvenile brownish, neck streaked. Unsociable ✦ Wetlands, mainly coastal, with abundant marsh or riparian vegetation (reedbeds, reedmace, tamarisks) where nests in small colonies among other herons; sometimes in trees. Feeds in marismas, lakes and ricefields ✦ Largely summer visitor ✦ Numbers (some 2,000 pairs, most of them in the Delta, la Albufera, the Alicante wetlands and the Marismas) fluctuate with water levels, but increasing and spreading. Scarce in winter (tens), chiefly in the Marismas, increasing. Scarce migrant in Balearics, but has nested latterly (Albufera de Mallorca). Rare and irregular in Canary Is. **NT**

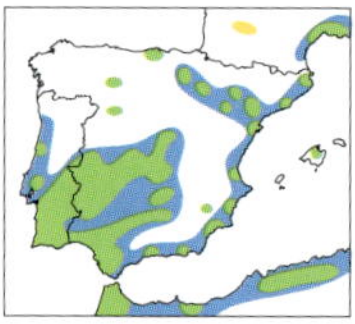

Cattle Egret *Bubulcus ibis* (garcilla bueyera). ~50 cm. Gregarious and often far from water. Small, chunky; neck, bill and legs comparatively short. Very white, with buff markings in nuptial plumage; bill and legs pale, reddish in nesting adults. Heavy jowl. Unlike Little Egret, legs protrude little in flight and neck retraction inconspicuous ✦ Nests colonially, usually near water, in trees, tamarisks or reedbeds; often alongside storks and other herons. Forages on pastures and croplands, including ricefields. Accompanies cattle and sheep; follows ploughs; visits rubbish dumps. Roosts sociably, often in large numbers ✦ Resident but highly dispersive when not nesting, often reaching as far as Morocco ✦ Population substantial (tens of thousands of pairs) but relatively concentrated: mainly in W Andalucía, Extremadura, Valencia and Delta. Declining in some regions but elsewhere increasing and spreading. Introduced to Cantabria. Scarce in Balearics (nests at Albufera de Mallorca, introduced). Appears regularly in Canary Is (nests Arrecife, Lanzarote).

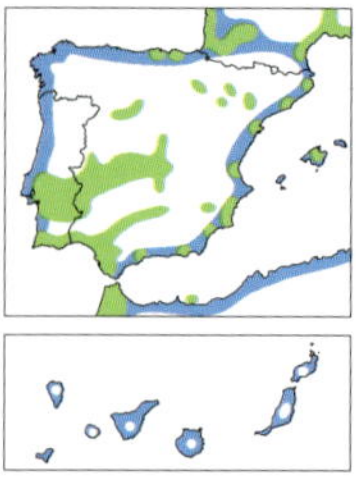

Little Egret *Egretta garzetta* (garceta común). 55–65 cm. Medium-sized, elegant and well proportioned. Bill and legs black; duller in juveniles. Feet yellow. Legs project prominently in flight, but less than in Great White Egret. Breeding birds have two long nape plumes. Occasional black or patchily dark individuals are vagrant **Western Reef-egrets** *Egretta gularis* (garceta dimorfa) or probable hybrids with that species, originating from W Africa. The Western Reef-egret occurs with some frequency at S and E wetlands, especially during Apr–Aug; adults very dark grey, nearly black, with throat white (p. 229) ✦ Nests colonially, among other herons; in trees, shrubs or marsh vegetation at lakes, rivers and reservoirs. Often also fishes in ricefields, saltpans and on the seashore ✦ Partial migrant; a good proportion overwinter ✦ Common (some 10,000 pairs, mainly at the Marismas, Albufera, Delta), rapidly increasing and spreading (now breeding on Biscay coast and in the Balearics and Chafarinas). Many thousands winter, chiefly at littoral sites in the E and S. Frequent on passage and in winter in Canary Is; has recently nested in Lanzarote.

ad. summer
Squacco Heron
Cattle Egret
ad. summer
Little Egret
ad. summer
Western
Reef-egret

Great White Egret *Ardea alba* (garceta grande). 85–100 cm. Relatively scarce but increasing. As large as Grey Heron, with even longer neck. Unlike Little Egret (previous page) has bill yellow (blackish when nesting), tibias yellowish, feet dark, loral skin yellowish or greenish, not bluish. Gape extends beyond eye. Flight ponderous, wingbeats slow, showing very prominent neck bulge and legs extending well beyond tail ✦ Littoral lagoons, ricefields, saltpans, reservoirs ✦ Present year-round, with largest numbers in winter and on passage (esp. Sept–Feb) ✦ Extinct breeder but recolonised Delta (1977) and Doñana (1978) and some tens of pairs now nest, at scattered sites (Delta, Marismas, Daimiel, Albufera de Mallorca), breeders clearly increasing. Many hundreds winter, largest concentrations in Catalonia and Albufera de Valencia. Vagrant to Canary Is.

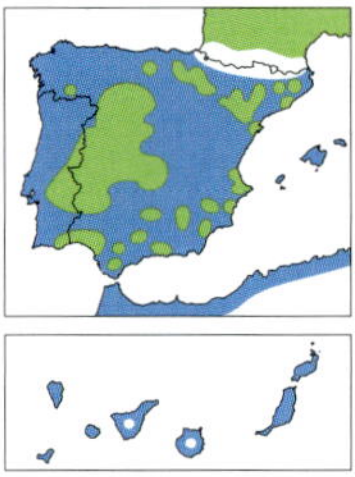

Grey Heron *Ardea cinerea* (garza real). 90–100 cm. The large common heron. Grey, with black and white head markings. Bill straight, stout. Breeding adult shows much white on head and neck, crown edged black, long black nape plume, bill yellow; immature more uniformly grey. Recalls Common Crane (p. 52) but neck retracted in flight. Flight deliberate, on arched wings. Call a harsh croak ✦ Habitat varied, especially in winter. Fishes in rivers, lakes, reservoirs, ricefields, saltpans and on coasts. Frequents tilled fields. Colonial nester, in trees (sometimes in reeds) ✦ Year-round, with wintering immigrants (Sept–Mar) ✦ Breeding population of several thousand pairs, rapidly increasing, concentrated especially in W (Duero, Tagus, Guadiana and Guadalquivir basins), but tending to spread elsewhere. Common (many thousands) and very widespread in winter, most frequent in littoral regions (especially abundant in Delta, Albufera and W Andalucía). Common in winter and on passage in Balearics (has bred) and Canary Is.

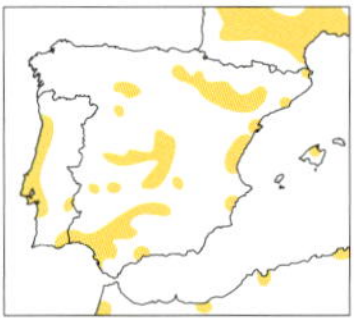

Purple Heron *Ardea purpurea* (garza imperial). 75–90 cm. Smaller, more graceful and darker than Grey Heron, with proportionately larger bill and feet; in flight shows narrower wings and a more prominent and angular neck bulge; call quieter and less resonant. Juvenile pale brown, with dorsal spotting and more muted neck striping than adult ✦ Associated with extensive wetlands and major rivers, seldom straying far from vegetation. Fishes in fresh waters. Nests in small loose colonies, within marsh vegetation (reedbeds, reedmace) ✦ Summer visitor, trans-Saharan migrant (Mar/Apr–Aug/Oct). Birds from elsewhere in Europe (France, The Netherlands) occur on passage. Very few winter records ✦ Quite common (a few thousand pairs, chiefly in littoral regions of S and E: Marismas, Delta, Albufera; scarcer inland: Extremadura, La Mancha, Ebro valley); some tens of pairs breed in Mallorca; increasing, with marked interannual fluctuations according to water levels. Range expansion has led to recent nesting locally at N coast wetlands. Scarce but regular in Canary Is.

ad. summer
non-breeding
Great White Egret
ad.
Grey Heron
ad.
Purple Heron

GANNETS AND BOOBIES (Suliformes, Sulidae) – Large seabirds that plunge-dive for fish from the air. Wings long and narrow; tail pointed, bill strong and large, lacks nares; legs short and feet totipalmate. Colonial nesters on islets and cliffs. Worldwide.

Northern Gannet *Morus bassanus* (alcatraz atlántico). 85–97 cm. Our largest seabird. Most often seen in flight, showing distinctive silhouette: cigar-shaped body and very long, narrow wings. Adults strikingly white but juveniles brown; immatures show increasingly whiter transitional plumages over 4 years. In flight alternates periods of wingbeats with short glides; may soar. When fishing may plunge-dive from a great height. Often in flocks, both when fishing or on passage, when forms long loose lines. Follows trawlers ✦ Uses coastal waters and readily seen from land ✦ In winter (Dec–Mar) and on passage (Sept–Nov and Mar–Apr) but many subadults especially remain in summer. Most originate from British Isles. Adults pass later in autumn and earlier in spring than younger birds ✦ Common (hundreds of thousands) off all coasts, including in the Canary Is, but relatively scarce in Balearic waters. Spectacular movements occur off northern headlands (Cape Bares, Cape Peñas). Many (tens of thousands) enter the Mediterranean through the Strait to winter.

CORMORANTS (Suliformes, Phalacrocoracidae) – Large fish-eaters, with mainly dark, metallic-glossed plumages. Float low in the water and surface-dive. Often rest sitting upright with wings extended. Bill hook-tipped; neck long; tail wedge-shaped; feet large, totipalmate. Juveniles and immatures paler, with whitish underparts. Worldwide.

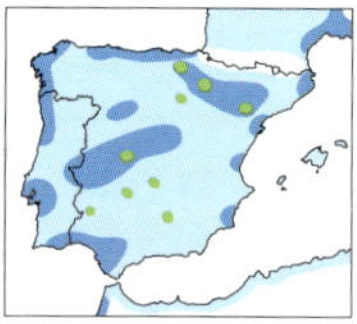

Great Cormorant *Phalacrocorax carbo* (cormorán grande). 80–95 cm. Only cormorant found inland. Robust, head large, bill and neck stout; gular area bare, yellowish, extends around eye. Flies with kinked neck; flight ponderous, with occasional glides. Breeding birds show pale areas on head and neck and white thigh patch ✦ Reservoirs, large rivers, estuaries, deltas, marshes, bays… Often in flocks. Rests and roosts on rocks, trees and wharves. Nests in trees ✦ Mainly winter visitor (passage: Sept–Nov and Feb–Apr), some remaining in spring and summer (nests Apr–Jul). Most originate from British Isles (ssp. *carbo*) and Low Countries, Denmark, Germany, etc. (ssp. *sinensis*) ✦ Rapidly increasing. Very common in winter (tens of thousands), with concentrations in the major river basins and in coastal regions of Catalonia, Valencia, Andalucía and Galicia; those on N coasts mainly *carbo*, elsewhere *sinensis*. Recently established breeder (ssp. *sinensis*), forming small colonies at reservoirs (hundreds of pairs, at Navalcán, La Serena, Giribaile and Castronuño reservoirs, etc.). Persecuted by fishermen. Rare in Canary Is.

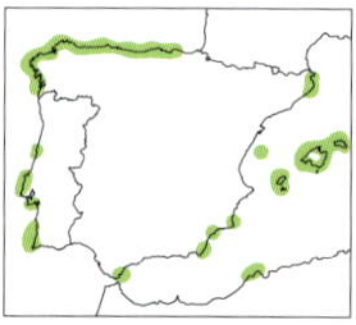

European Shag *Phalacrocorax aristotelis* (cormorán moñudo). 65–80 cm. Strictly marine. Confusable with Great Cormorant at long range but smaller, more slender, with thin beak, steep forehead, dark throat and yellow gape flanges; flight swift, with neck extended, close to the water and without glides. Breeding birds have black crest at first (short in ssp. *desmarestii*) ✦ Rocky coasts and sea cliffs (often on islands and islets) where forms normally small colonies, nesting on ledges or in small caves. Often fishes in tight groups ✦ Sedentary but dispersive, sometimes far-ranging. Nests early (from Feb in the Mediterranean and Mar/Apr in N) ✦ Ssp. *aristotelis*: Atlantic, quite common (hundreds of pairs, majority on Cíes and Ons islands, Galicia), declining; **EN**. Ssp. *desmarestii*: Mediterranean some 2,000 pp. in Balearics and a few tens of pairs on mainland, increasing; **VU**. Vagrant to Canary Is.

Northern Gannet
3rd year
ad. winter
juv.
ad. non-breeding
Great Cormorant
breeding
juv.
breeding
European Shag

OYSTERCATCHERS (Charadriiformes, Haematopodidae) – Large black or pied coastal waders, with long, flattened bills. Feed on bivalves and other invertebrates. Cosmopolitan.

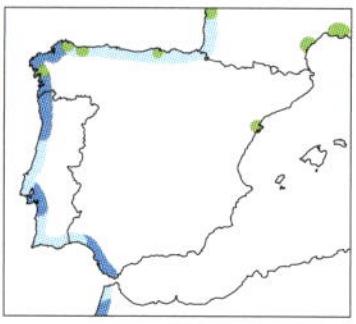

Eurasian Oystercatcher *Haematopus ostralegus* (ostrero euroasiático). 40–45 cm. Unmistakable. Large, plumage highly contrasted, bill red, legs pink. Has white throat band in winter. Juvenile and 1st winter brownish above, with greyish legs and dark bill tip; off-black back and dark bill tip persist into 1st summer. In flight shows broad white wingbar, white rump and black terminal tailbar. Noisy, even at night; a repeated piping whistle *kleep*. The extinct, similar but all-black **Canarian Black Oystercatcher** *Haematopus meadewaldoi* (ostrero negro canario) inhabited rocky coasts in the Canary Is, in Fuerteventura, Lanzarote and associated islets; it survived until at least 1913 ✦ Chiefly estuaries and beaches with a broad intertidal zone. Nests on coastal islets (N) and beaches with dunes (Delta) ✦ Chiefly on passage and in winter (Jul/Oct–Mar/May), from the British Isles, North Sea and Scandinavia. Some oversummer ✦ Scarce breeder (a few tens of pairs) and very local (sites in Galicia, Asturias and Cantabria, and Delta); **NT**. A couple of thousand winter, chiefly in Galicia and W Andalucía; scarce in the Mediterranean and occasional inland. Rare in Balearics and scarce in Canary Is.

STILTS AND AVOCETS (Charadriiformes, Recurvirostridae) – Elegant pied shorebirds with very long legs and long straight or curved bills. Hind toe absent or vestigial. Eat insects and other invertebrates. Nest colonially. Hot or warm regions worldwide.

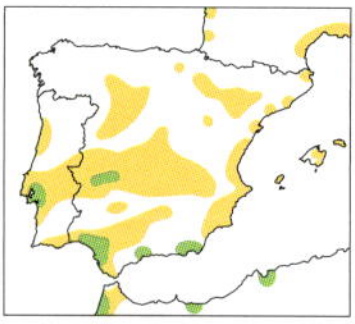

Black-winged Stilt *Himantopus himantopus* (cigüeñuela común). 35–40 cm. Elegant and delicate, with extremely long pink legs and very fine bill. White with black wings and mantle; head and hindneck variably black in adults; browner in winter and in juveniles. Juveniles have pale feather edging above and white trailing edge to wings. Black glossier in males. Incessantly clamorous when alarmed: *kreet–kreet–kreet–kreet...* ✦ Shallow fresh or brackish waters: ricefields, marismas, estuaries, reservoirs, gravelpits, saltpans, lakes and seasonal pools ✦ Summer visitor and migrant (Feb/May–Jul/Oct), but many winter ✦ Common (many thousands of pairs) and locally abundant (Marismas, Delta; Valencia, Alicante and Murcia wetlands, Cádiz Bay, Odiel estuary). Increasing, especially inland, with large fluctuations according to water levels; more widespread in seasons when the Marismas are dry. Winterers concentrated in S (thousands), increasing. Mainly scarce migrant in N, where sometimes nests. Common in Balearics. Regular migrant in Canary Is, has nested Lanzarote and Gran Canaria.

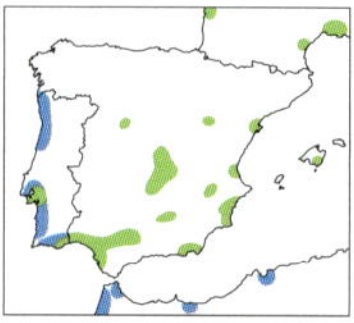

Pied Avocet *Recurvirostra avosetta* (avoceta común). 42–45 cm. As elegant as it is unmistakable, with its black and white plumage and upcurved bill. Feeds while wading, sweeping bill from side to side. Often swims. Female has duller head markings and somewhat shorter and more sharply upcurved bill. Juvenile shows brownish markings above. Metallic piping call: *kluit, kluit, kluit* ✦ Shallow fresh or brackish waters, mainly coastal (saltpans, marismas, lakes) ✦ Breeding birds partly migratory; passage of birds from The Netherlands and nearby countries (Aug/Oct–Mar/May) ✦ Breeding birds fairly common (a few thousand pairs) but very concentrated (Marismas, Cádiz Bay, Delta, Santa Pola) and relatively scarce inland (La Mancha, Villafáfila); increasing. Thousands winter, especially in Andalucía. Scarce in Balearics; nests locally in Mallorca. Scarce but regular in Canary Is.

Eurasian Oystercatcher
ad. winter
ad. summer
Black-winged Stilt
♀
♂
Pied Avocet

PLOVERS AND LAPWINGS (Charadriiformes, Charadriidae) – Small or medium-sized shorebirds. Often run. Relatively large-headed, with large eyes and short bills. Catch invertebrates, using 'watch, run, pick' foraging. Ground nesters. Worldwide.

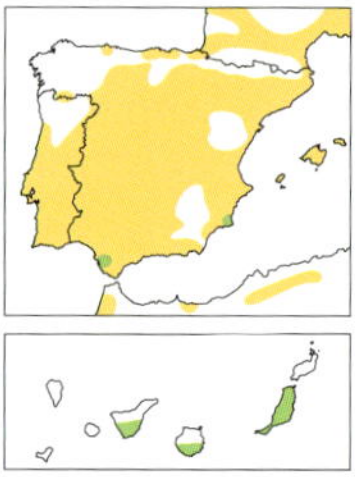

Little Ringed Plover *Charadrius dubius* (chorlitejo chico). ~15 cm. Complete collar permits confusion with Common Ringed Plover but is smaller, slimmer, lacks wingbar, has pale legs and bill is almost entirely black. Call: *pi–u,* slightly downslurred. Adults have conspicuous yellow eye-ring and black forehead bar fringed white above; ear coverts browner in females. Winter adults have browner head and collar markings. Juvenile paler, with pale feather edging above (lacks white supercilium of juvenile Common Ringed) ✦ Nests by fresh water at gravelpits, sandpits, riverine sandbanks, pool and lake shores… At other times visits estuaries, marismas, ricefields, saltpans ✦ Summer visitor and migrant (Mar/Apr–Jul/Oct), some winter ✦ Ssp. *curonicus*: breeds fairly commonly (a few thousand, perhaps increasing) and widely, especially inland. Scarce in winter (tens or a few hundreds) chiefly on S and E coastlands. Nest in Balearics (tens). Scarce and local nester in Canary Is (tens) and occurs on passage on many islands.

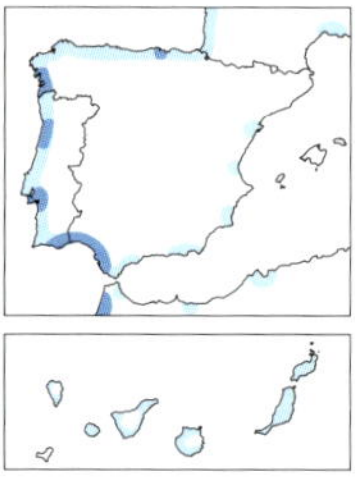

Common Ringed Plover *Charadrius hiaticula* (chorlitejo grande). ~18 cm. Common. Compact, with complete collar and conspicuous wingbar. Call distinct from Little Ringed Plover: a melodious, disyllabic, rising *too–eep*. Breeding adult has striking orange bill with black tip, bright orange legs and broad black forehead band. Black areas become brownish in winter, bill and legs darken, and collar may become incomplete centrally. Juvenile similar to winter adult, with pale feather edging above ✦ Chiefly coastal, on intertidal mudflats (estuaries, beaches) and saltpans; also on shores of pools and lakes inland ✦ Passage and wintering (Mar/May–Aug/Oct) of birds from N and W Europe, Iceland and Greenland; a few oversummer ✦ Quite common winterer (a few thousand), especially in W Andalucía. Common and widespread on passage on coasts, especially Atlantic; scarce inland. Common in Balearic and Canary Is.

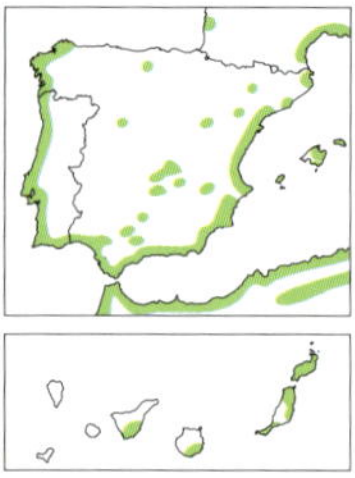

Kentish Plover *Charadrius alexandrinus* (chorlitejo patinegro). 15–17 cm. Intermediate in size between above two species, differing in blackish legs and incomplete collar. Appears pale, sandy; legs long and set well back, allowing characteristic rapid running. Calls: a quiet *pit–pit–pit*, and *prrr* when alarmed. Breeding male has variably rusty nape, black forehead bar and broad white supercilium; female duller. Juvenile shows pale feather edging above. Nests in loose colonies ✦ Partial migrant, present year-round. Passage and wintering of birds from France, The Netherlands, Germany, etc. (mainly Aug–Oct and Mar–Apr) ✦ Nests on coasts (saltflats, quiet sandy beaches, saltpans) and also on 'steppe' lagoon shores inland ✦ Fairly common breeder (a few thousand), declining, high concentrated on E littoral (Delta and Alicante, Murcia and Almería wetlands) and in S (Cádiz bay, Marismas del Guadalquivir and Odiel estuary). Present but scarce in Galicia and inland (Fuentedepiedra, La Mancha, Aragón). Sparse migrant on Biscay coast. Winters (a few thousand), especially in W Andalucía. Common in Balearics. Common in Canary Is (a few hundred; eastern islands). **VU**

ad. winter
Little Ringed Plover
ad. summer
ad. summer
ad. winter
Common Ringed Plover
♂ ad. summer
Kentish Plover
ad. winter

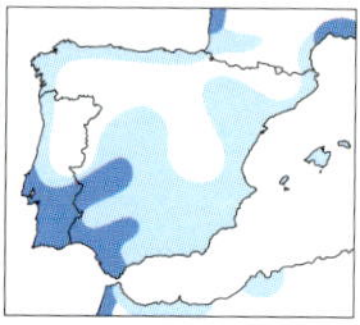

Eurasian Golden Plover *Pluvialis apricaria* (chorlito dorado europeo). ~27 cm. Mainly inland. Winter adults and juveniles have golden-toned cryptic plumage. Unlike Grey Plover has white axillaries and indistinct wingbar. Call: a liquid, whistled *luu–eee*. Flight rapid, often in tight flocks; recalling sandgrouse. Often with Lapwings. The **American** *P. dominica* (chorlito dorado americano) and **Pacific** *P. fulva* (chorlito dorado asiático) **Golden Plovers**, rarities (p. 229), are smaller and slimmer, with obvious white supercilia and primary projections longer than tail; tertials longer than tail in Pacific ✦ Pastures and arable land, alfalfa. Visits estuaries, ricefields, saltpans ✦ Wintering and passage (Oct–Nov and Mar–Apr) of birds from Iceland, British Isles and Scandinavia ✦ Tens of thousands in winter, esp. in SW (Extremadura, W Andalucía). Common in Balearics; scarce and irregular in Canary Is.

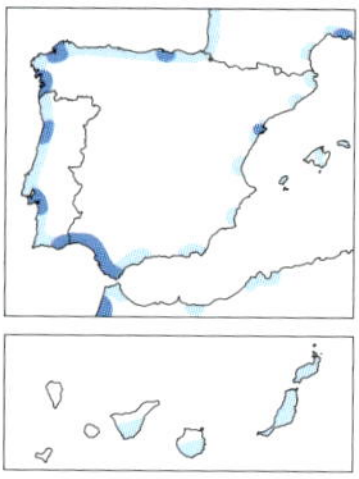

Grey Plover *Pluvialis squatarola* (chorlito gris). ~28 cm. Coastal. Sturdier than Golden Plover, with larger head and bill, obvious wingbar and diagnostic black axillaries. Adults greyish but juveniles have yellowish-buff tones. Call different; a plaintive trisyllabic *peee-uu-eee*. Hindclaw present ✦ Intertidal mudflats, estuaries, beaches, saltpans ✦ On passage and in winter (Aug/Oct–Apr/May); some oversummer. Nests at very high latitudes, the nearest in N Russia ✦ Some 6,000 winter, most on Atlantic coasts (Cádiz Bay, Marismas del Guadalquivir and Odiel, Rías Bajas). Common on passage on all coasts; uncommon inland. Scarce in Balearics; common in Canary Is.

Eurasian Dotterel *Eudromias morinellus* (chorlito carambolo). ~22 cm. Recalls Golden Plover but smaller, has long supercilia that meet at nape, pectoral band and shorter, yellowish legs. Breeding plumage striking, esp. in females, but in Spain most seen are moulting adults or juveniles; latter with indistinct breastband and darker above with broad pale feather edgings. Wings unmarked except for white shaft of outermost primary. In small groups. Confiding ✦ Nests on Pyrenean plateaus above 2,400 m. Migrants stage on mountaintops, arable land and stubbles ✦ Passage (Aug–Oct and Apr–May) of Scandinavian birds that winter in N Africa ✦ Has bred on rare occasions in La Cerdaña. Scarce autumn migrant, rarer in spring, at widespread regular staging sites. Has overwintered on semiarid plains of Almería. Rare in Balearics and Canary Is. **EN**

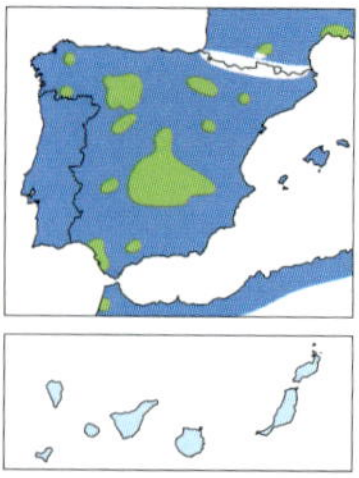

Northern Lapwing *Vanellus vanellus* (avefría europea). ~30 cm. Well known in winter. Easily identified: pied, with distinctive crest, rounded wings, floppy flight and ringing *pee-wee* call. In breeding birds, female has shorter crest and some white on throat; male has broader 'hand'. Winter adult has face mottled buff and black; dorsal feathers edged buff. Juvenile very short-crested, with faint facial markings and pale fringing above. Sociable, tends to nest in small groups. Display flight acrobatic ✦ Nests on lake margins, marismas, sedgefields, wet meadows. Also winters on arable land, ricefields, around reservoirs ✦ Resident, with many wintering, passage (Oct–Dec and Feb–Mar) and summering birds from C and N Europe. Massive influxes with cold snaps ✦ Scarce nester (1,000–1,500 pairs), fluctuating, most often in Castilla and W Andalucía (Villafáfila, Daimiel, Marismas). Tens or hundreds of thousands in winter, very variable, most abundant in SW and littoral regions. Very common in Balearics and scarce but regular in Canary Is.

Eurasian Golden Plover
ad. summer
winter
winter
Grey Plover
ad. summer
Eurasian Dotterel
♀ summer
winter
winter
♂ summer
Northern Lapwing

STONE-CURLEWS (Charadriiformes, Burhinidae) – Medium-sized, cursorial birds, generally in steppe and semidesert environments. Plumage cryptic, eyes very large, tarso-tibial joint swollen; toes short, hind toe absent. Sexes similar. Of crepuscular or nocturnal habits. Chiefly insectivorous. Widely distributed worldwide.

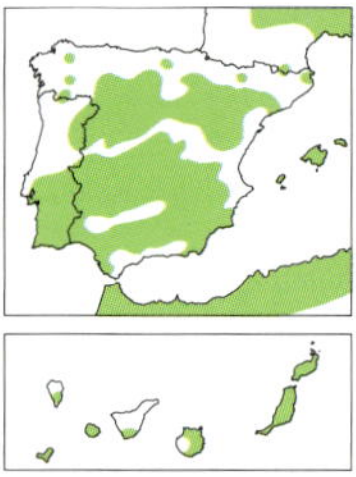

Eurasian Thick-knee *Burhinus oedicnemus* (alcaraván común). 40–45 cm. Lanky, with large yellow eyes; legs and bill base also yellow. Relatively short-necked, with long wings and tail. White, black-edged wingbar, most prominent in male. In flight shows second wingbar and white primary patches. Juvenile largely lacks supercilium and shows reddish edging to scapulars and upperwing coverts. Runs and freezes, staring fixedly. Wailing *currr–liii* calls, esp. at dusk. Flocks in autumn and winter ✦ Open country with low vegetation; fallows, dry pastures, crops (cereals, sunflowers, vines), sandflats, firebreaks, stony plains ✦ Partial migrant. Some French and English birds occur on passage and in winter (Sept/Oct–Mar/Apr) ✦ Ssp. *oedicnemus*: Mainland, common (several tens of thousands of pairs) and widespread except in humid N and mountains; declining. Ssp. *saharae*: Balearics, common and widespread. Ssp. *insularum*: common in Lanzarote and very common in Fuerteventura, **EN**. Ssp. *distinctus*: Other Canary Is, rare and declining, **EN**

SANDPIPERS AND ALLIES (Charadriiformes, Scolopacidae) – Includes most waders. Sizes, and relative proportions of bills and legs, highly variable. Seek buried invertebrates. Many nest at high latitudes and are highly migratory. Cosmopolitan.

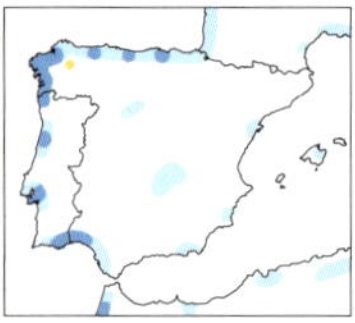

Eurasian Curlew *Numenius arquata* (zarapito real). 50–55 cm. Our largest wader; size variable, females largest. Bill very long; markedly and uniformly decurved. Rump white. Large wings and relatively slow wingbeats suggest a large gull in flight. Outer primaries and primary coverts contrast darker. Juvenile has 'fresher' plumage; upperpart feathers brightly contrasting, dark-centred, pale-fringed. Flight call a distinctive, rising *cour–lee* ✦ Mainly coastal, on intertidal mudflats, estuaries, bays, beaches ✦ In winter and on passage (Jul–Oct and Feb–Apr), some oversummer. Nests in C and N Europe ✦ A very few pairs nest in N (Terra Chá, Galicia, formerly in Asturias); **EN**. Common on passage; some thousands in winter, chiefly on Atlantic coasts (Santoña, Galicia, Marismas del Odiel). Less common (hundreds) in the Mediterranean (Delta); rare inland; scarce in Balearics. Scarce to fairly common in Canary Is.

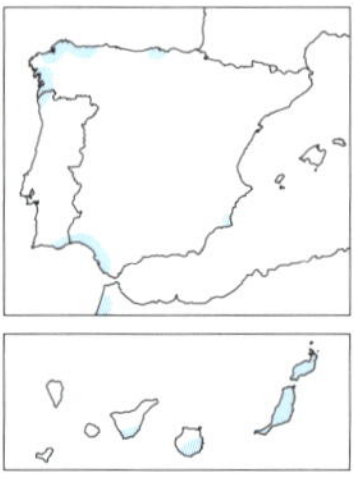

Whimbrel *Numenius phaeopus* (zarapito trinador). ~40 cm. Smaller, shorter-billed and shorter-legged than Curlew; head striped; dark cap with pale median line. Shorter-winged and wingbeats faster than Curlew. Flight pattern recalls Bar-tailed Godwit; tail-barred and rump white. Call distinctive: a seven-note whistle ✦ Rocky, muddy or sandy shores ✦ Migrant (Mar–May and Jul–Sept), with some winterers and scarce oversummering birds. Nests Iceland and N Europe ✦ Common, especially on Atlantic coasts. Uncommon in winter (a few hundred), chiefly in Galicia, Cantabria and W Andalucía. Rare inland on passage. Scarce in Balearics. Common in Canary Is, on passage and in winter.

Eurasian Thick-knee

Eurasian Curlew

Whimbrel

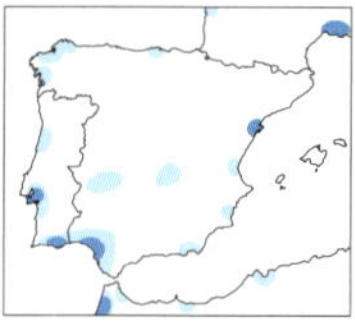

Black-tailed Godwit *Limosa limosa* (aguja colinegra). ~40 cm. Tall; neck, legs and bill long, nearly straight. Very different from Bar-tailed Godwit in flight, showing bold white wingbar and black and white tail. Face, neck and breast variably rufous in breeding plumage, especially in males, but belly white, barred. In winter uniformly grey except for white belly; unstreaked above, unlike Bar-tailed. Juvenile with neck and breast orange-buff, mantle and scapulars dark brown, fringed buff. Often feeds in quite deep water. Flocks, often in large numbers ✦ Chiefly littoral, in marismas, coastal lagoons, saltpans and estuaries, but also found inland, especially on ricefields ✦ In winter and on passage (Jun–Oct and Jan–Mar), some oversummering. Ssps. *limosa* (C Europe and Russia) and, fewer, *islandica* (Iceland, Scotland and N Norway) ✦ Common on passage, both on coasts and inland, but generally declining (increasing in Extremadura). Some tens of thousands winter (W Andalucía, Delta, Albufera), but arrivals on northward passage apparent as early as January. Has bred or attempted to breed sporadically at various sites; **VU**. Scarce in Balearics. Uncommon but regular in Canary Is.

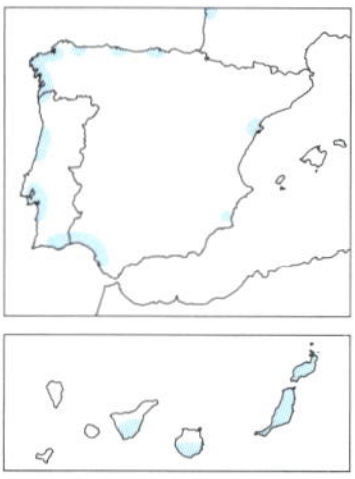

Bar-tailed Godwit *Limosa lapponica* (aguja colipinta). ~38 cm. Resembles Black-tailed Godwit but flight pattern recalls curlews: lacks wingbar; tail white, finely barred; rump and lower back white. Unlike Black-tailed is shorter-legged, esp. tibias; bill distinctively upturned; female longer-billed and longer-legged than male. Browner than Black-tailed in winter, clearly streaked above, supercilium longer. Breeding plumage male rufous, including belly; female pinkish buff below; juvenile buff below, with pale feather edging above ✦ Highly maritime, unlike Black-tailed: intertidal mudflats, rías, estuaries, bays, beaches ✦ Migrant (Aug–Oct and Apr–Jun) with some wintering; a few individuals oversummer. Nests in N Scandinavia and Siberia, winters chiefly in Mauritania ✦ Common on Atlantic coasts (Cádiz bay, Guadalquivir and Odiel marismas, Galician rías, Santoña), scarce in Mediterranean. A few hundred winter, chiefly in W Andalucía and Galicia. Scarce in Balearics. Common in Canary Is.

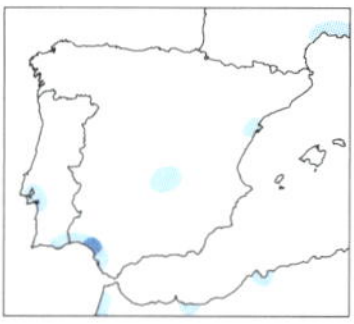

Ruff *Calidris pugnax* (combatiente). 22–32 cm. Males some 20% larger than female. Male breeding plumage is flamboyant but birds in Spain usually less spectacular. Somewhat intermediate between shanks and sandpipers but distinctive. Bill longish, tip slightly downturned; head small; neck long; hunched looking at times; long mantle feathers often raised; legs variably orange or greenish. In flight shows fine wingbar and white oval areas fringing rump. In winter, head grey, streaked, ofen white around bill base. Juvenile has face and breast buff, upperpart feathers broadly fringed cream. Flies lazily, with glides. Nearly mute. Often in flocks ✦ Muddy areas, chiefly freshwater: flooded fields, ricefields, lakes, pools ✦ On passage and in winter (Jul/Oct–Feb/May); originate from The Netherlands, Poland and Scandinavia ✦ Common on passage, especially in spring, both coastally and inland. Some hundreds winter, increasing (Marismas, Delta, irrigation zones of Extremadura, La Mancha). Common in Balearics. Scarce but regular in Canary Is.

Black-tailed Godwit
winter
♂ summer
winter
♂ summer
Bar-tailed Godwit
♂
♀
winter
Ruff
juv.

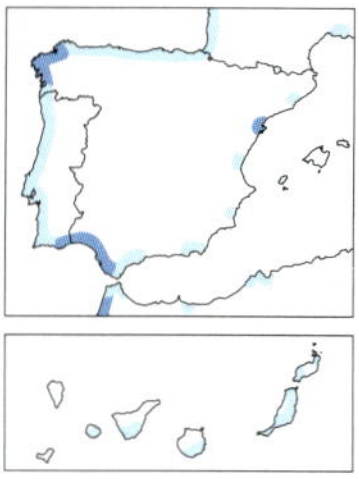

Sanderling *Calidris alba* (correlimos tridáctilo). ~20 cm. Fast runner at water's edge. Very pale, bill and legs black. Lacks hindclaw. Adult winter shows black carpal patch, more evident in flight when shows white, black-bordered wingbar, more obvious than in other small waders. In breeding plumage head, breast and upperparts variably chestnut with black markings, but belly remains strikingly white. Juvenile chequered black and white above, cap dark, necksides black and cream. Call: *twick–twick* ✦ Sandy beaches; sometimes on estuaries, mudflats… ✦ On passage and in winter (esp. Aug/Oct–Mar/May), of Arctic origin; some oversummer ✦ Common or very common on coasts, especially on the Atlantic; occasional inland. Some thousands winter on sandy shores (Cádiz Bay, Rías Bajas, Delta). Uncommon in Balearics. Common in Canary Is, particularly in autumn and on eastern islands.

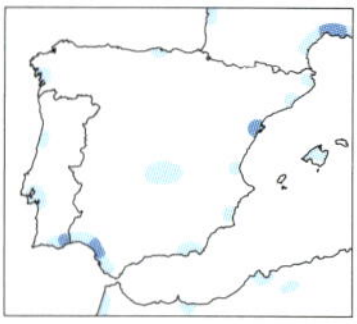

Little Stint *Calidris minuta* (correlimos menudo). ~14 cm. Tiny. Bill short and straight (unlike Dunlin) and legs black (unlike Temminck's Stint). In flight shows thin white wingbar and dark-centred tail edged grey (white in Temminck's). In breeding plumage brownish red above, marked black; dark cap, mantle edges make white V. Quite uniformly greyish in winter, like many small waders. Juvenile attractive; mantle V prominent and mantle and scapulars dark-centred, edged rufous. Energetic. Flight pattern as Dunlin. Call: a squeaky *stit*, sometimes repeated ✦ Coastal or inland, on muddy areas or pools at lakes, coastal lagoons, estuaries, ricefields, saltpans ✦ On passage (Jul–Oct and Mar–Jun) and in winter ✦ Common on passage (more so on the Atlantic side in autumn and in the Mediterranean in spring). Relatively scarce (a few thousand) in winter (mainly in S and E: Marismas, Delta, Cádiz Bay, Alicante wetlands, La Mancha). Common in Balearics. Scarce but regular in Canary Is.

Temminck's Stint *Calidris temminckii* (correlimos de Temminck). ~14 cm. As tiny as Little Stint but legs yellowish and very short; bill slightly decurved, body elongated, tail edges white. Brownish in summer, with darker scapulars. Juvenile shows typical pale-fringed upperparts. Discreet, creeping about in low vegetation; easily overlooked. Call: a sharp trilling *tirririr* ✦ Inland or on coast. Prefers fresh waters: lakes, pools, river mouths ✦ Migrant (esp. Aug–Oct and Apr–May), some winter; nests eastwards from Scandinavia and N Russia ✦ Scarce or under-recorded on passage, both on coasts (chiefly Mediterranean) and inland. A few winter, mainly in S and E. Scarce in Balearics. Rare in Canary Is.

Purple Sandpiper *Calidris maritima* (correlimos oscuro). ~21 cm. On rocky coasts. Easily identified; dark grey, with short, yellow-orange legs. Longish, slightly decurved bill has yellowish base. Largish, short-necked, dumpy. Head and upperparts uniform greyish in winter. Looks paler in summer, with black, rufous and white upperpart markings. Juvenile with white feather edgings above. Flies low; hops about rocks and swims freely. Sometimes very confiding. Call: *twit* or *twit–wit* ✦ Seashores, on rocks, islets, breakwaters ✦ Winters (Sept/Oct–Apr/May). Nests in boreal latitudes from Scandinavia and Iceland onwards ✦ Uncommon or scarce (hundreds), perhaps declining, on Biscay coasts and in N Galicia. A few reach Andalucía. Occasional or vagrant elsewhere, including the Balearic and Canary Is.

Sanderling
winter
Little Stint
winter
Temminck's Stint
winter
Purple Sandpiper

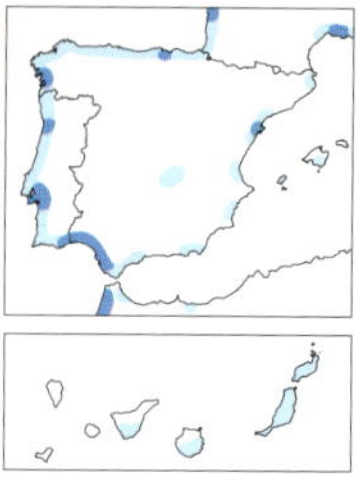

Dunlin *Calidris alpina* (correlimos común). 17–20 cm. Abundant. The baseline typical wader for identification purposes. Size and relative bill length very variable, the latter often longer than the head. Summer adults distinctive with black belly patch but nondescript grey in winter. Juvenile has very streaked breast and dense black spotting on flanks. Bill gently decurved. Legs black. Narrow white wingbar. Rump and tail dark-centred, rump sides and outer uppertail coverts white. Call: *treer* or *treep*. Sociable, flies in tight flocks ✦ Intertidal mudflats in estuaries and inlets; also marismas, ricefields, saltpans, lakes ✦ In winter and on passage (mainly Aug–Sept and Apr–May), some oversummer. Subspecies occurring are *alpina* (large and long-billed; breeding in N Scandinavia and NW Russia), *schinzii* (Iceland, British Isles and Baltic region) and *arctica* (short-billed, from NE Greenland, scarce) ✦ Very common on passage on all coasts, especially in autumn. Very common in winter (tens of thousands, in Delta, Cádiz Bay, Marismas, Odiel estuary, Galician rías, Santoña, etc.). Scarce inland. Regular on passage and in winter in Canary Is.

Curlew Sandpiper *Calidris ferruginea* (correlimos zarapitín). ~19 cm. Elegant, long-necked; bill longer than head and clearly decurved (hence 'curlew'). Legs black, relatively long. Uppertail coverts and lower rump conspicuously white. Summer adult strikingly brick red with upperparts marked rufous, black and grey. Greyish in winter, with prominent supercilium. Often in intermediate plumage in summer/autumn. Juvenile shows scaly pattern above and creamy-buff neck and breast. Call: a soft *chir-rup* ✦ Diverse flooded areas: saltpans, lagoon shallows, estuaries… ✦ Migrant (mainly late Jul–Sept and Apr–May). Nests in N Siberia, wintering in Africa ✦ Fairly common; particularly in autumn on Mediterranean coast. A few hundred winter, nearly all in W Andalucía (Marismas, Cádiz bay). Quite common on passage in Balearics. Regular on passage in Canary Is, where a few winter.

Red Knot *Calidris canutus* (correlimos gordo). ~24 cm. Large and stout. Bill rather short; legs relatively short, greenish or greyish. Tail and rump grey. Underparts brick-red in summer but greyish in winter. Juvenile has scaly appearance above; feathers neatly edged dark subterminally with white margins; also buffish below and shows noticeable supercilium. Call: *knut* ✦ Coastal, on intertidal mudflats, estuaries, beaches ✦ On passage and in winter (Aug/Oct–Apr/Jun), originating from N Canada, N Greenland and N Siberia (many travelling to/from Mauritania) ✦ Uncommon on Atlantic shores and scarce in the Mediterranean, but sometimes in large flocks. Scarce winterer (hundreds or a few thousand), most often in Galicia and W Andalucía. Scarce but regular in Balearics and Canary Is.

Pectoral Sandpiper *Calidris melanotos* (correlimos pectoral). 19–23 cm. Rare. Size of Green Sandpiper. Neatly streaked breast contrasts sharply and abruptly with white belly. Bill shortish, slightly decurved, paler at base; legs pale; yellowish. Juvenile shows white mantle and scapular Vs; dark upperpart feathers neatly fringed white and cream, giving scaly effect. In flight shows indistinct wingbar, dark-centred tail and white sides to rump and uppertail coverts. Call: *trrrt–trrrt* ✦ Diverse wetlands, both freshwater and brackish ✦ Migrant (late Jul–Oct and Apr–May), originating from N Canada or NW Siberia ✦ Very scarce in autumn (tens) and exceptional in spring; most observed at coastal wetlands of Galicia/Asturias and Catalonia/Valencia, seldom inland; juveniles predominate, in Sept–Oct. Rare in Balearics. Very scarce in Canary Is.

Dunlin
winter
winter
Curlew Sandpiper
winter
Red Knot
winter
Pectoral Sandpiper

Sandpipers. Juvenile plumage

Sandpipers and Ruff. Nuptial plumage

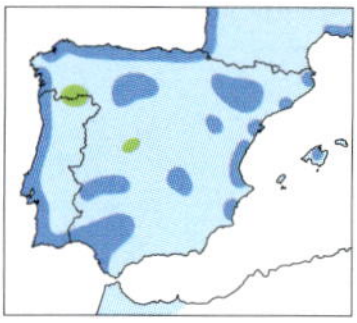

Common Snipe *Gallinago gallinago* (agachadiza común). 25–27 cm. Bill extremely long, at least twice head length, very obvious even in flight. Streaked, flanks barred, belly white, legs short. Has pale median crown stripe and broad buff supercilium. When flushed flies high in a zig-zag pattern, giving characteristic harsh, rasping call: *scaaap*. Tail then shows very little white. Feeds in soft ground and shallow water, probing rapidly with the bill, 'sewing machine' fashion. External rectrices produce a curious drumming sound during display flights ✦ Flooded areas, especially vegetated ones: wet meadows and fields, marismas, ricefields, pools, rivers, streams ✦ Chiefly on passage (Aug/Nov and Feb/Apr) and in winter, originating from C and N Europe (ssp. *gallinago*) and perhaps Iceland (*faeroeensis*) ✦ Ssp. *gallinago*: very scarce and local breeder (a few tens of pairs, in Gredos, S Galicia and perhaps elsewhere in north); **EN**. Very common on passage and in winter (many tens of thousands), especially in littoral regions; routinely hunted. Regular but scarce in Canary Is.

Great Snipe *Gallinago media* (agachadiza real). 26–30 cm. Rare. Somewhat larger and bulkier than Common Snipe and shorter-billed. Shows much white on outer tail feathers when flushed and a prominent wingbar in flight. Usually flushes silently and flies off low and straight ✦ Flooded areas in meadows and grasslands ✦ On passage (chiefly Mar–May and Sept–Oct) between NE Europe and Africa ✦ Formerly considered a rarity; most often found in coastal Catalonia and the Balearics, in spring.

Jack Snipe *Lymnocryptes minimus* (agachadiza chica). 18–20 cm. Some 30% smaller than Common Snipe; bill barely as long as head. Shows two conspicuous yellow stripes on each side of the dark upperparts; median crown stripe dark; supercilium pale with a small dark central bar resembling an eyebrow. Flushes silently at close range, flying only a short distance, slowly and low, before dropping into cover with wings raised. Skulking, like a crake, and most easily seen from hides ✦ Wet, muddy, well-vegetated areas: meadows with sedgebeds, reedbed margins, marismas, ricefields ✦ On passage and in winter (Oct/Nov–Feb/Apr), from Scandinavia, Finland, N Russia ✦ Scarce (a few thousand) probably declining. Widespread, perhaps most frequent in littoral regions. Hunted incidentally. Scarce in Balearics and vagrant to Canary Is. **DD**

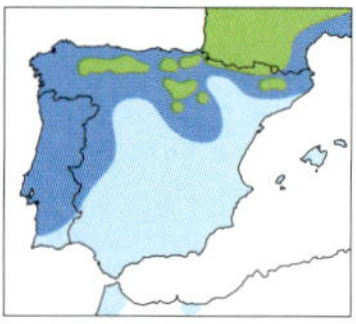

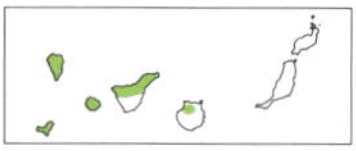

Eurasian Woodcock *Scolopax rusticola* (chocha perdiz). ~35 cm. A woodland species, very hard to see. Bill long (one-and-a-half times head length). Brownish-rufous plumage is highy cryptic, with broad transverse barring on the crown and dense barring below. Stouter and more compact than Common Snipe; wings relatively short and broad. Takes flight noisily when flushed, twisting through the trees. Usually solitary; feeds at night in open areas. Male has a crepuscular aerial display in spring, the roding flight, flying slowly above the tree canopy along an habitual route, making several croaking sounds followed by a sneezing note ✦ Nests in humid montane woodlands (of beech, oak, birch, pine). Also winters in dehesas and open scrub woodland ✦ Partial migrant, breeding birds perhaps resident but joined by European birds on passage and in winter (Nov/Dec–Jan/Feb), these sometimes arriving in great numbers during cold snaps ✦ Breeding population small (a few thousand), in mountains of N and Sistema Ibérico. Also nest in western Canary Is, especially La Gomera (particularly in laurel forest). Common and widespread in winter, especially in the north and in cold winters (hundreds of thousands); Common in Balearics. Much hunted.

Common Snipe
Great Snipe
Jack Snipe
Eurasian Woodcock

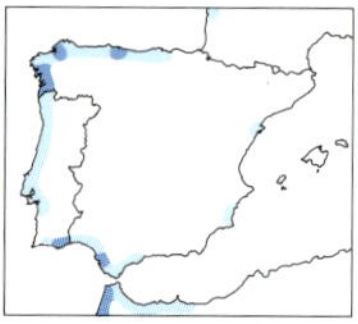

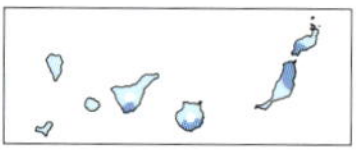

Ruddy Turnstone *Arenaria interpres* (vuelvepiedras común). ~23 cm. Unmistakable. Dumpy, with short orange legs, a short, conical bill, and a black or blackish pectoral band contrasting with the very white belly. Pied flight pattern is striking, including white back. Nuptial plumage showy, especially in males, with black and white head pattern and brownish orange scapulars and coverts. Upperparts brownish-grey in winter. Juvenile has fine cream or ginger feather edges to scapulars, coverts and tertials. Most frequent call a rapid, low-pitched, *tuk–a–tuk–tuk*. Usually in small flocks ✦ Seashores, particularly on pebbly beaches or those with seaweed, intertidal rocks, breakwaters, spits, saltpan walls ✦ On passage and in winter (Aug/Sept–Apr/May). Originates from NE Canada, Greenland, Scandinavia, N Russia ✦ Common or very common on passage on Atlantic coasts, including in the Canary Is; uncommon or scarce in the Mediterranean. Scarce in the Balearics. Occasional inland (most often in spring). A few hundred winter, chiefly on Biscay coasts and in Galicia, W Andalucía and the Canary Is.

Red-necked Phalarope *Phalaropus lobatus* (falaropo picofino). ~18 cm. Phalaropes are small, confiding waders that nest in boreal or arctic latitudes and winter at sea, feeding there on plankton. They swim habitually, often in tight circles to draw food to the surface. Nuptial plumages, especially of females, are colourful but in winter they are grey above and white below, with characteristic black ear covert patches. The Red-necked Phalarope has a needle-shaped, all-black bill; it is smaller and shorter-winged than the Red Phalarope; in winter it shows contrasting white edges to the upperpart feathers and only an indistinct black mark on the hindcrown. Juveniles are brownish-black above with ochre striping and pinkish-buff necksides. **Wilson's Phalarope** *Phalaropus tricolor* (falaropo tricolor), a vagrant from N America (p. 231), lacks black ear patches and has a very long neck and even longer bill ✦ Usually appears on saltpans or littoral lagoons ✦ On passage (mainly Jul–Oct and Apr–May). Breeds Iceland, Scotland, Scandinavia, Finland, Russia... apparently crosses Eurasia to winter in the Arabian Sea (but a Scottish bird was tracked to Ecuador/Perú) ✦ Annual, most frequent in autumn; a few, usually solitary, individuals, on S and E coasts (esp. the Delta). Very rare or vagrant inland and in the N. Vagrant to the Balearics and perhaps also the Canary Is.

Red (Grey) Phalarope *Phalaropus fulicarius* (falaropo picogrueso). ~21 cm. Larger and stockier than Red-necked Phalarope, with thicker, less pointed bill. Uniformly grey in winter, with an obvious black hindcrown patch and often some yellow at the bill base. Juvenile dark above, lacking the ochre striping of the Red-necked above; soon moulting to grey 1st winter plumage ✦ Pelagic, seldom on coasts but may turn up in rías, inlets, saltpans, tidal pools and, occasionally, on inland lakes ✦ Autumn migrant (Aug–Nov), with some in winter (Dec–Jan/Feb). Nests in the Arctic and winters off coastal W Africa ✦ Flocks regular and quite common on passage off NW coasts in autumn (Cape Peñas, Cape Bares), scarcer in SW Andalucía. Storm-driven birds occasionally appear on coasts, often as tideline corpses, and may then occur exceptionally in the Mediterranean or inland. Spring birds (Apr–May) extremely rare. Vagrant to Canary Is.

Ruddy Turnstone
♂ ad. summer
♂ ad. summer
♀ ad. summer
juv.
Red-necked Phalarope
Wilson's Phalarope
♀ ad. summer
Red (Grey) Phalarope
ad. winter

Spotted Redshank *Tringa erythropus* (archibebe oscuro). ~30 cm. Red-legged, like Common Redshank, but lacks white in wing. More or less black in breeding plumage, when unmistakable; grey, and the palest 'shank, in winter. Slim; bill long and fine; straight with slightly drooping tip, some red at base of lower mandible; lores black, white supercilium prominent. In flight shows white oval dorsal patch and barred tail. Juvenile darker above and below than winter adult (and juvenile Common Redshank), densely and uniformly barred below. Call: *chu–wit*. Often in small flocks. Wades quite deeply; swims readily ✦ Prefers fresh waters: estuaries, ricefields, pools; but also on saltpans ✦ On passage (Mar–May and Jul–Oct) and in winter. Nests from N Scandinavia to Siberia ✦ Quite common on passage, on coasts and inland. A few hundred winter, chiefly in Galicia, Marismas and Delta. Scarce in Balearics and Canary Is.

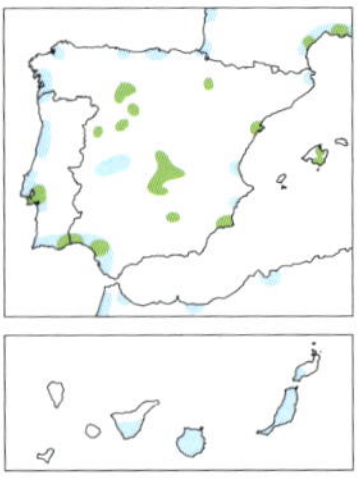

Common Redshank *Tringa totanus* (archibebe común). ~27 cm. Common. Legs red or orange; in flight white secondary and inner primary patch diagnostic. Also shows white dorsal wedge. Well-proportioned; bill straight, reddish, dark-tipped. Plumage brownish and unremarkable, with much black spotting in summer; more uniform greyish in winter. Juvenile has duller legs and is darker above, with buffish feather edging. Noisy but melodious; the familiar flight call is a ringing *teu–tu–tu*; in nuptial flight calls *tyoo* repeatedly ✦ Nests in wetlands among low vegetation: in marismas, lagoon margins, wet pastures. At other times chiefly coastal: in estuaries, bays, saltpans, saltmarshes, ricefields ✦ Partial migrant. European birds occur on passage (Jul–Oct and Mar–May), in winter and oversummer ✦ Common on passage and in winter (some thousands), chiefly in W Andalucía, Delta and Galicia; Common in Balearics and Canary Is. Few breed (several hundred), most in the Marismas; **VU**.

Marsh Sandpiper *Tringa stagnatilis* (archibebe fino). ~23 cm. Quite rare. Delicate, very long-legged, long-necked, bill very long and fine. Resembles a miniature Greenshank in its greenish legs and flight pattern but bill much finer. Dark carpal joint area noticeable at rest. Brownish-grey above and white below in winter; black-spotted in summer. Juvenile shows pale feather fringes above. Calls: *tew*, and a sharp *chip* in alarm ✦ Shallow flooded areas in marismas, coastal lagoons, ricefields ✦ On passage (Mar–Apr and Jul–Oct) and in winter ✦ Scarce migrant, most frequent in spring, on the Mediterranean littoral and in the Balearics. Some (tens) winter in the Marismas. Rare inland. Vagrant in the N and Canary Is.

Common Greenshank *Tringa nebularia* (archibebe claro). ~32 cm. Large; legs greenish. Bill long, stout-based, bluish grey, tip dark and slightly uptilted. Shows long white dorsal wedge in flight. In summer shows some black dorsal feathers and clearly streaked neck, breast and flanks. Juvenile darker above, feathers variably showing pale fringes and dark edge spotting. Call very characteristic: a ringing *teu–teu–teu*. Active and energetic ✦ Diverse wetlands, frequently coastal: rías, estuaries, bays, intertidal mudflats. Sometimes catches small fish ✦ On passage and in winter (Jul/Oct–Mar/May), some oversummer. Nests beyond N Scotland and S Norway ✦ Quite common on coasts on passage, less frequent inland. Uncommon in winter (a few hundred), in Galicia, Marismas, Cádiz bay, Delta. Scarce in Balearics. Fairly common in the Canary Is (E islands especially).

Spotted Redshank
summer
winter
Common Redshank
Marsh Sandpiper
summer
winter
Common Greenshank
winter
summer

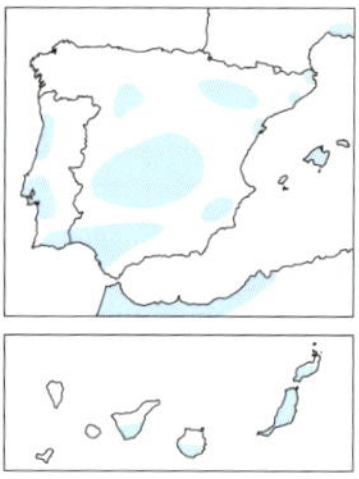

Green Sandpiper *Tringa ochropus* (andarríos grande). ~22 cm. Dark, stocky, wings uniformly blackish above and below, contrasting with white rump. Dark upperparts and dark breast sharply divided horizontally from white underparts. Both straight bill and legs medium-long, legs barely projecting beyond tail in flight. White tail shows a few broad transverse bars. Orbital ring white; supercilium only extends forwards from eye. Legs greenish. Wing coverts and tertials show some white marginal spotting that declines through wear by summer. Juvenile has brownish-grey head and breast, less streaked than adult. Takes flight with loud call: *tlueet–wit–wit*. Often solitary ✦ Diverse fresh waters, including very small ones: pools, streams, irrigation channels ✦ On passage and in winter (mainly Jul/Sept–Mar/May). Breeds N Europe, onwards from Germany and Poland ✦ Quite common on passage and relatively widespread and common in winter (a few thousand), both inland along major river valleys and in littoral areas. Common in Balearics. Scarce but regular in Canary Is.

Wood Sandpiper *Tringa glareola* (andarríos bastardo). ~20 cm. Quite slim, with long neck and legs; recalls a small redshank rather than a sandpiper. Rump white as in Green Sandpiper but wings brown above and very pale below; legs project clearly beyond tail in flight. Prominent supercilium extends behind eye. Tail finely barred. Legs yellowish. Upperparts coarsely spotted white. Juvenile browner above, with cream spotting. Breeding birds show barred flanks. Call: a sharp, rapid *chiff–if–if* ✦ Muddy shorelines or small pools at freshwater wetlands ✦ Migrant (Mar–May and Jul–Sept). Breeds in Scandinavia, Finland, Russia ✦ Common on eastern littoral (Catalonia, Valencia, Balearics), scarce or rare elsewhere. Scarce but regular migrant in Canary Is. Scarce winterer (SW Andalucía, Delta).

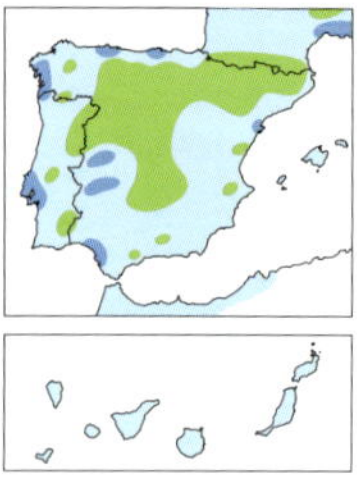

Common Sandpiper *Actitis hypoleucos* (andarríos chico). ~20 cm. Unusual shorebird, having short legs and bill and a long tail that projects beyond wingtips (giving an elongated appearance). Continually teeters, rocking rear end up and down. Flight characteristic; at low levels, erratic wingbeats on downward arched wings alternating with short glides. White wingbar conspicuous; feet do not project beyond tail in flight. Brownish above; white of underparts extends up breast sides ahead of wings. Juvenile shows obvious barring above and buff mottling on tips and edges of tertials and wing coverts. Flight call: *swee–wee–wee*. The **Spotted Sandpiper** *Actitis macularius* (andarríos maculado), a N American vagrant (p. 231), is very similar in non-breeding plumage but has shorter tail projection, yellower legs, shorter wingbar and a different call. The **Terek Sandpiper** *Xenus cinereus* (andarríos del Terek), a very rare migrant from NE Europe and Asia (p. 231), has yellow legs and an upturned bill ✦ Nests alongside clean rivers, especially on spits and gravel banks. At other times more widespread at wetlands, and also on rocky coasts, rías and harbours ✦ Partial migrant, with winterers, migrants (Jul–Oct and Mar–May) and oversummering birds from C and N Europe ✦ Scarce breeder (perhaps some hundreds of pairs; many non-breeders present in summer), most frequent in N and mountainous regions, rare in warmer and flatter areas. Widespread and very common on passage and less numerous (hundreds) in winter. Common in Balearics. Regular on passage and fairly common in winter in Canary Is.

Green Sandpiper
ad. summer
ad. summer
Wood Sandpiper
ad. winter
ad. summer
Common Sandpiper

BUTTONQUAILS (Charadriiformes, Turnicidae) – Small, quail-like terrestrial birds of Africa, S Asia and Australia. Like quails, they skulk in herbaceous vegetation. Formerly included in order Gruiformes. Plumages dun, highly cryptic. Legs short, lack hind digit; wings short and tail nearly vestigial. Females larger and more showy than males, the latter solely caring for clutches and the nidifugous young. Ground nesters.

Common Buttonquail *Turnix sylvaticus* (torillo andaluz). ~16 cm. Extreme S, extremely rare. Hardly detectable except by the nocturnal mooing song of the female, which gives it its Spanish name 'little bull': a slow, resonant *mooooo–mooooo–mooooo…*, recalling blowing through a tube. Somewhat smaller than the Common Quail, whose coloration it recalls, but has a plain orange breast and triangular black flank markings. Male duller ✦ Low scrub (sage-leaved cistus, lentisc, dwarf fan palm) with dense grass ✦ Sedentary ✦ Very scarce (a few individuals), confined to a few sites in coastal Huelva and Cádiz, perhaps now extinct. Common until the early 20th century in SW Andalucía; tiny numbers survive in Morocco (however, widespread in sub-Saharan Africa and S Asia). **CR**

COURSERS AND PRATINCOLES (Charadriiformes, Glareolidae) – Small or medium-sized cursorial Old World (chiefly African) birds, bills short and decurved, wings long and pointed. Leg length variable, long in coursers and shorter in pratincoles. Insectivorous. Sexes similar. Ground nesters.

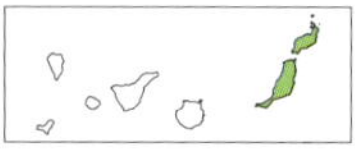

Cream-coloured Courser *Cursorius cursor* (corredor sahariano). ~24 cm. Canary Is. Small, slim and plover-like; sandy-coloured with long, very pale legs. Shortish bill decurved, eyes large, broad supercilia meet on nape. Adult has black eyestripe below white supercilium, rusty forehead and blue-grey crown. Juvenile and first-winter show indistinct head pattern; juvenile spotted on breast and above. In flight shows black outerwing above and all-black underwing. Runs rapidly, standing very erect when pausing. Flight rapid. Flight call: *praak–praak* ✦ Arid stony or sandy plains, with scattered low vegetation. Eat beetles, grasshoppers and other invertebrates ✦ Resident but nomadic (partial migrant on African continent) ✦ A couple of hundred pairs, perhaps declining; common in Fuerteventura, scarce and local in Lanzarote and now apparently extinct in Gran Canaria; **EN**. From Africa, occasionally reaches the Peninsula (p. 232) where has bred in Almería and Albacete provinces, and presumably also Tenerife and Gran Canaria.

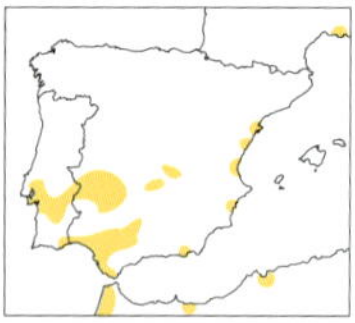

Collared Pratincole *Glareola pratincola* (canastera común). ~25 cm. Long-winged and deeply fork-tailed; in flight resembles a marsh tern or large swallow. Flies in noisy flocks, calling *kikki-kirrick, kikki-kirrick* harshly. Very cryptic when settled, suggesting a 'partridge-headed' plover. Juvenile lacks bib and shows dark markings on breast and above. Nests in noisy colonies ✦ Open plains, often arid or saline, frequently near water (peripheries of marismas, lakes and ricefields, also on pastures, fallows and tilled land, preferably with pools or irrigation reservoirs). Catches insects in flight ✦ Summer visitor (Mar/May–Jul/Sept) ✦ Uncommon (a few thousand, fluctuating), chiefly along lower Guadalquivir (Sevilla, Marismas) but also in Extremadura, La Mancha and the Levant coast. Scarce migrant in Balearics. Scarce but regular in Canary Is. **VU**

Common Buttonquail

Cream-coloured Courser

Collared Pratincole

GULLS AND TERNS (Charadriiformes, Laridae) – Long-winged birds of both coastal and inland waters, with predominantly white and grey plumages. Gulls consume fish and garbage. Terns take fish and insects, typically caught in flight. Colonial nesters, usually on the ground although marsh terns nest on floating vegetation. Young semi-nidifugous. Variably migratory, some travelling extremely long distances. Worldwide.

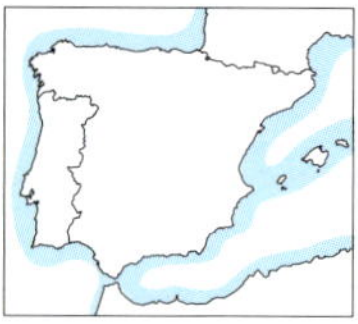

Little Gull *Hydrocoloeus minutus* (gaviota enana). ~26 cm. Tiny, recalls a marsh tern (but tail unforked). Underwings of adults black or blackish; upperwings grey with narrow white rear margin. Juvenile has blackish mantle but soon moults into the grey of 1st winter plumage; dark primaries and median coverts make a W-pattern above, recalling immature Kittiwake, but is much smaller, has dark cap and lacks black half-collar. 1st summer shows a partial hood. Calls tern-like ✦ Uncommon (hundreds or a few thousand). Most often noted in autumn (Aug–Dec) on Biscay coasts and in Galicia and in spring (Mar–May) in Andalucía and the Mediterranean. Rare inland. Scarce in Balearics. Occasional in Canary Is.

Sabine's Gull *Xema sabini* (gaviota de Sabine). ~30 cm. Pelagic, seldom observed inshore. Small, with forked tail and characteristic wing pattern: white innerwing triangles contrast with black outer primaries and grey mantle and coverts. Adult retains trace of nuptial hood. 1st winter has brown wing coverts and black tail fork, unlike similar but larger 1st winter Kittiwake ✦ Seldom approaches land except when storm-driven ✦ Summer/autumn migrant (Aug–Nov). Breeds in Greenland, N Canada, Alaska, Siberia ✦ Nearly all records are from Asturias and Galicia, where regular offshore in autumn; occasional in winter and spring. Vagrant to Canary Is.

Black-legged Kittiwake *Rissa tridactyla* (gaviota tridáctila). ~40 cm. Pelagic. May suggest a Common Gull, but has short black legs and solid black 'dipped in ink' wingtips. Tail slightly notched. Juvenile and 1st winter resemble those of Little and Sabine's Gulls but are larger and have conspicuous black half-collar ✦ Come to land solely to nest, on sea cliffs, or when storm-driven ✦ Year round. Wintering and passage (Oct–Dec and Feb–Apr) of birds from British Isles, France, Norway, etc. ✦ Nested in very small numbers on Galician coast (Cape Vilán, Sisargas Is), where first established in 1970s; recently extinct; **VU**. Common offshore in winter in Biscay and Atlantic waters; uncommon or rare in the Mediterranean. Occasional inland after storms. Scarce and irregular in Balearic and Canary Is.

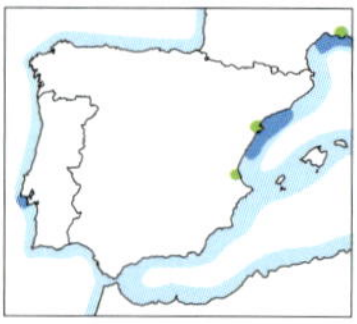

Mediterranean Gull *Larus melanocephalus* (gaviota cabecinegra). ~38 cm. Adult is only gull with all-white flight feathers. Slightly larger and stockier than Black-headed Gull, with rounder, broader wings; bill shorter and thicker; in winter auricular mark more diffuse and extensive. 2nd winter usually shows small black primary tips. 1st winter has black terminal tail bar, black primary wedge and brown median coverts, contrasting with grey midwing panel; recalls 1st winter Common Gull but has dark auricular mask and dark legs ✦ Mainly coastal; follows trawlers ✦ Winters (Sept/Nov–Feb/Mar), some present in other months. Largest breeding concentrations in Black Sea ✦ Common on passage and in winter in the Mediterranean (tens of thousands, increasing) and much less common (hundreds or a few thousand, increasing) on Biscay coasts and in Galicia; uncommon inland. Small numbers have nested recently (Delta, Albufera, Santa Pola). Scarce in Balearics. Occasional in Canary Is.

Little Gull
winter
1st winter
summer
Sabine's Gull
summer
1st summer
winter
Black-legged Kittiwake
ad. summer
winter
juv.
Mediterranean Gull
summer
1st winter
1st winter
winter

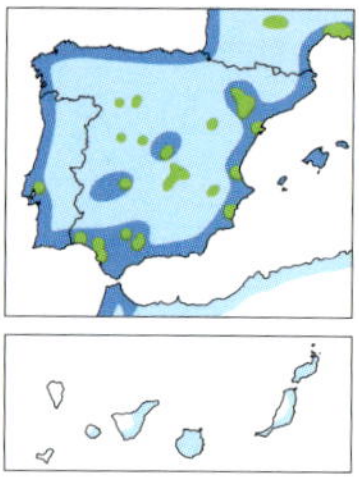

Black-headed Gull *Larus ridibundus* (gaviota reidora). ~36 cm. Abundant. Upperwing shows a white flash on the outer primaries and primary coverts, as in Slender-billed Gull; underside of primaries, except the outermost, grey. In nuptial plumage (Feb–Aug) has dark chocolate facemask, not a complete hood as in Little and Mediterranean Gulls. Bill red, nearly black in summer. 1st winter shows brownish median coverts and a black tail band. Juvenile has brown cap, neck and upperparts. Harsh calls distinctive; colonies noisy. Often in large flocks, flying in V-formation ✦ Nests in small colonies on lakes, marismas, saltpans. Opportunistic; feeding in coastal waters as well as in ricefields, rivers, rubbish dumps; follows plough ✦ Partial migrant and winterer, with passage (Jul–Nov and Feb–Apr) and over-summering of European birds ✦ Very common in winter (several hundred thousand), formerly chiefly on coasts but increasingly inland (especially Madrid). Breeding population (a few thousand, increasing) of recent origin (post 1960), most concentrated in E and S. Common in Balearic and Canary Is.

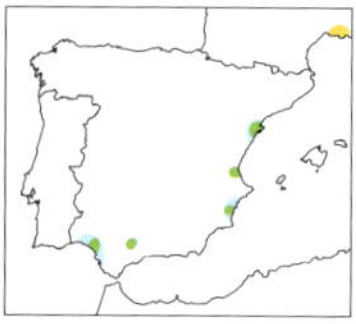

Slender-billed Gull *Larus genei* (gaviota picofina). ~42 cm. E and S coasts. Resembles Black-headed Gull, with similar wing pattern, but lacks nuptial hood and is slimmer, distinctively long-necked and long-billed (not slender-billed); iris pale. Breeding adult has pink-tinted underparts. 1st winter resembles that of Black-headed Gull but has only slight ear spot, yellowish-brown legs and bill, latter with little or no black on tip ✦ Colonial nester at saltpans or brackish marismas ✦ Mainly summer visitor (Mar–Aug), but some numbers winter ✦ Scarce (around one thousand pairs), local (most in Marismas and Delta; other colonies in Valencia and Almería). Exceptional inland and in N. Rare in Balearic and Canary Is. **VU**

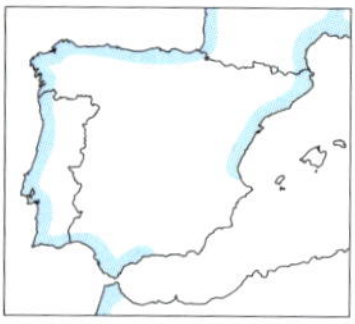

Mew Gull *Larus canus* (gaviota cana). 40–45 cm. Winters in N. Medium-sized; bill and legs yellowish, head rounded, bill smallish; eyes dark; large white outer primary mirrors striking. 1st winter recalls that of Mediterranean Gull but lacks auricular mask and has pale legs and bill base. 2nd winter as adult but shows black on primary coverts and a black bill-ring. Confusable with N American **Ring-billed Gull** *Larus delawarensis* (gaviota de Delaware) (p. 232), a very rare winterer in NW; which is larger, stronger-billed and has pale iris when adult ✦ Chiefly coastal ✦ Winters (Aug–Jun, peaking Jan), commoner in cold years ✦ Scarce in Galicia and Biscay coast; very scarce in Mediterranean and W Andalucía; occasional inland, especially during cold snaps. Rare in Balearics. Occasional in Canary Is.

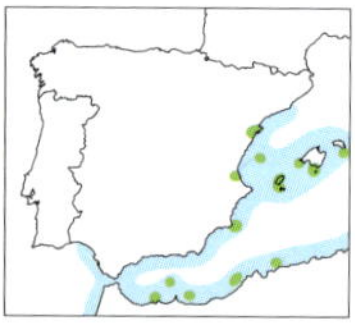

Audouin's Gull *Larus audouinii* (gaviota de Audouin). ~50 cm. Mediterranean. Considerably smaller than Yellow-legged Gull, more delicate and elegant, with paler grey upperparts, grey extending to neck. Forehead elongated; drooping bill dark red; legs grey; eyes dark. Juvenile darker than juvenile Yellow-legged, with grey legs, pale bill base, head brownish with pale crown, mantle and scapulars dark with contrasting pale feather edging; tail black with U-shaped white uppertail covert patch. 1st winter paler overall, with greyer mantle and scapulars; 2nd year birds grey above but remiges are dark; tail has narrow dark terminal bar. 3rd years as adults, often with some black on primary coverts ✦ Coastal. Nests on islands, saltpans, etc. Follows trawlers, often at night ✦ Partial migrant, most travelling to W and N African coasts (passage ~Jul–Oct and Feb–May) ✦ Common (some 20,000 pairs), increasing, formerly concentrated at Delta and Chafarinas Is, but now also has large colonies in Castellón and La Mata-Torrevieja. Widepread and common in Balearics. Occasional in N and inland. Occasional in Canary Is. **VU**

Black-headed Gull
summer
1st winter
winter
summer
1st winter
winter
Slender-billed Gull
ad. winter
1st winter
Mew Gull
ad. summer
ad. summer
Audouin's Gull

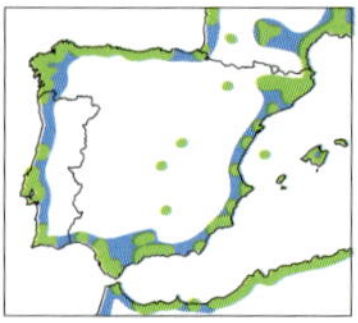

Yellow-legged Gull *Larus michahellis* (gaviota patiamarilla). ~60 cm. Very common on coasts. Confusable with other species on this page, the **Caspian Gull** *Larus cachinnans* (gaviota del Caspio) and other occasional gull species (p. 232). Adults have legs yellow; grey upperparts contrast with black wingtips, orbital ring red; head barely streaked in winter. Juvenile and 1st winter have pinkish legs; resemble those of other large gulls but have very pale head, contrasting dark remiges, and a neat terminal tail bar. 2nd winter greyer above with pale bill base. 3rd winter as adult with residual dark marking on bill, wings and tail ✦ Nests on islets, sea cliffs, sometimes on rooftops. Follows trawlers and feeds at harbours, rubbish dumps, etc. ✦ Present year round, with displacements ✦ Ssp. *michahellis*: abundant (many tens of thousands, increasing) and widespread, although most numerous in Galicia, Biscay coasts, Catalonia, Balearics and Chafarinas. Uncommon inland, where some nest at reservoirs. Ssp. *atlantis*: Canary Is, common (a few thousand, increasing).

European Herring Gull *Larus argentatus* (gaviota argéntea europea). ~60 cm. Scarce winterer in N. Adult has pink legs; compared with Yellow-legged Gull is paler grey above; wingtips show less black and have larger white mirrors; orbital ring yellowish. Head and neck heavily streaked in winter. 1st winter shows contrasting pale wedge on inner primaries; greater coverts densely barred but not darker than rest of wing (unlike 1st winter Yellow-legged); tail bar contrasts poorly with barred uppertail coverts; bill base pale ✦ Coastal ✦ Wintering (~Oct–Feb) of birds from France, England and nearby countries (ssp. *argenteus*) and perhaps Scandinavia (*argentatus*) ✦ Uncommon in Biscay coastlands and scarce or rare on other coasts. Vagrant to Balearic and Canary Is.

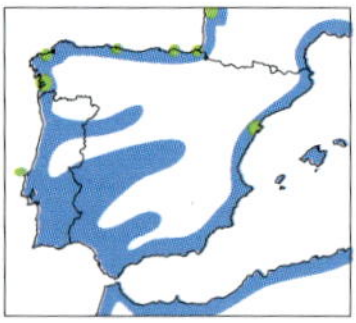

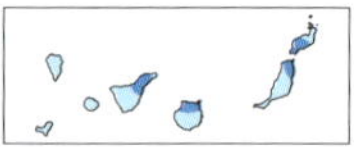

Lesser Black-backed Gull *Larus fuscus* (gaviota sombría). ~55 cm. Common in winter. Legs yellow. Smaller than Yellow-legged Gull and darker above; wingtips poorly contrasted. Head and neck heavily streaked in winter. 1st winter lacks pale primary wedge of 1st winter Herring Gull and is darker than that of Yellow-legged Gull, with unbarred tertials ✦ Coastal when nesting; also inland in winter, when attends rubbish dumps and roosts on reservoirs ✦ Chiefly in winter and on passage (~Aug–Dec and Feb–Apr), some over-summer. Most are ssp. *graellsii* (British Isles, France), with some *intermedius* (Netherlands, Denmark, Norway) and rare *fuscus* (Baltic and Finland) ✦ Very common in winter (several hundred thousand) particularly in S Galicia, Gulf of Cádiz and Alborán basin; also abundant, increasingly, inland (Andalucía, Extremadura, Madrid). Abundant on passage on Atlantic coasts. Breeding population of recent origin (hundreds, increasing), chiefly in Galicia (*graellsii*) and the Delta (*intermedius*). Scarce in Balearics. Common in Canary Is.

Great Black-backed Gull *Larus marinus* (gavión atlántico). ~70 cm. Very large (although some may be smaller than some Yellow-legged Gulls). Adults have black wings and mantle and pale flesh-pink legs. 1st winter very like Yellow-legged Gull, with white head and black bill, but bill is very robust with a very pronounced gonys; dark centres to upperpart feathers give check pattern effect; tertials have broad white tips and there is a marked tertial step ✦ Coastal ✦ In winter (~ Sept–Feb), with some at other times. Breeds in France, British Isles, Norway ✦ A few pairs now breed in NW. Otherwise scarce in Galicia and Biscay coastlands (Rías Bajas), rare in W Andalucía; occasional in the Mediterranean. Vagrant to Balearic Is. Scarce and irregular in Canary Is.

juv.
Yellow-legged Gull
ad.
European Herring Gull
ad. winter
ad.
Lesser Black-backed Gull
ad. summer
Great Black-backed Gull

Identifying Gulls

Identifying **adult** gulls is not usually too difficult, given that upperwing colours, wingtip patterns and the colours of bills and feet are often distinctive. Younger birds, generally brownish, are another matter, particularly because the diversity of immature plumage often causes confusion. It is therefore important to understand the moult sequence. The first complete plumage, in which the young bird fledges, is the **juvenile** plumage. This is followed later the same year by a partial moult into the **first winter** plumage, replacing wing coverts but not the remiges and rectrices. There is a further springtime partial moult into a **first summer** plumage in some species. Thereafter a sequence of complete annual moults in summer-autumn and partial moults towards early spring results in a series of immature winter and summer plumages. The wing coverts and body feathers thus tend to be replaced twice a year but the primaries and tail only once, so that these latter often look heavily worn in summer. The complete moult that follows the first summer results in the **second winter** plumage. This is identical to the adult winter plumage in many small species, such as the Black-headed Gull. Such species are termed **two-year gulls** since they acquire adult plumage in the calendar year following the hatching year. In **three-year gulls**, such as the Kittiwake, the second winter plumage differs from the adult plumage, as does the second summer plumage. Many gulls, such as Audouin's Gull and all the larger species, are **four-year gulls**, which develop both third winter and third summer plumages, during which they typically retain some immature-type wing and tail feathers and often have duller bills than full adults, often with dark markings.

Audouin's Gull

Black-legged Kittiwake

Black-headed Gull

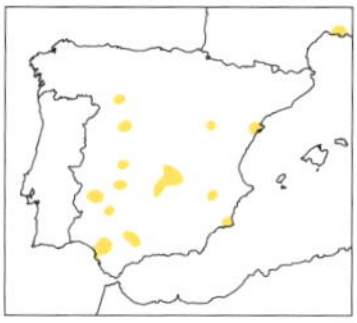

Common Gull-billed Tern *Gelochelidon nilotica* (pagaza piconegra). ~36 cm. Not maritime. Bill all black, quite short and thick, gull-like. Legs black, somewhat longer than in other terns. Rump and tail grey, tail shortish and slightly forked. More compact and broader-winged than other terns; flight deliberate. Outer primaries have dark trailing edge. Head white in winter plumage with dark eye mask; this less marked in 1st years. Juvenile has pale brownish crown and more faintly patterned above than other terns. Flight call: *kawack*, distinctive. Flocks circle over dry farmland ✦ Nests colonially, often with other species, in emergent sites in marismas, lakes, reservoirs, saltpans. Hawks insects over fallows, pastures, crops ✦ In summer and on passage (Apr–May and Aug–Sept) ✦ Uncommon (a few thousand), increasing, on plains, both littoral (Marismas, Delta, Albufera) and inland (Guadalquivir valley, Extremadura, La Mancha). Passage sparse but also evident in N. Scarce in Balearics. Irregular and scarce in Canary Is. **VU**

Caspian Tern *Hydroprogne caspia* (pagaza piquirroja). ~50 cm. Massive, with a striking, thick, dark-tipped, red bill. The **Royal Tern** *Thalasseus maximus* (charrán real), a vagrant from Mauritania or N America (p. 233), has orange bill without dark tip and is 20% smaller and more slender. Undersides of primaries dark. Head large; tail short and slightly forked. 1st winter has darker outer primaries and dark tail markings. Call harsh, deep: a heron-like *kraa* ✦ Coastal lakes, bays, estuaries ✦ Wintering and passage (Aug–Oct and Mar–Apr) of birds nesting in the Baltic ✦ A few dozen winter in SW Andalucía. Scarce on passage in the Mediterranean and even scarcer in N. Vagrant inland and in the Balearic and Canary Is.

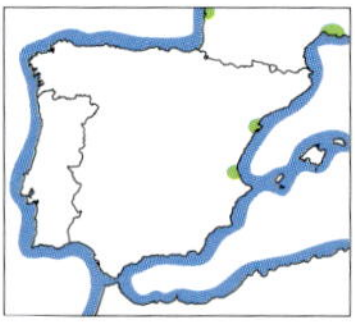

Sandwich Tern *Thalasseus sandvicensis* (charrán patinegro). ~40 cm. Coastal. Largish, slender. Bill long and slender, black with yellow tip. Legs black. Black cap ends in shaggy crest; forehead white in winter. Very pale; outer primaries darker. Juvenile has bold scaly markings above; retains juvenile tertials and dark tail tips in 1st winter. Call: a sharp *kirrik*. Patrols shores and plunges from a height ✦ Nests on islets in coastal lagoons and saltpans, foraging on the coast ✦ Year-round, with passage of birds of Atlantic (France, British Isles, Netherlands) and Black Sea origins, especially Mar–May and Aug–Nov ✦ Breeds locally (a few thousand), increasing, at the Delta (since 1970s) and Albufera (since 1990s), with attempts elsewhere. Common in winter (a few thousand), mainly in Mediterranean (Delta), SW Andalucía and Rías Bajas. Common on passage. Fairly common in Balearics. Common in Canary Is. **NT**

Lesser Crested Tern *Thalasseus bengalensis* (charrán bengalí). ~36 cm. Strait region. Resembles Sandwich Tern but darker grey above, somewhat slighter in build and longer-winged; bill orange, slender. Juvenile and first winter have dark wing and tail-fork markings. **Royal Tern** (see above) has similar bill colour but is considerably larger and paler ✦ Coastal ✦ On passage (around May and Jul–Nov). Nests in Libya (ssp. *emigrata*) and winters in W Africa ✦ Very scarce in the Strait and Alborán basin, more frequent on Maghreb coast; exceptional elsewhere. Most records in autumn. Has bred at the Delta and Albufera (isolated pairs, sometimes mixed with Sandwich Tern). Vagrant to Balearic and Canary Is.

Common Gull-billed Tern
winter
summer
Caspian Tern
winter
juv.
Sandwich Tern
juv.
Lesser Crested Tern
winter

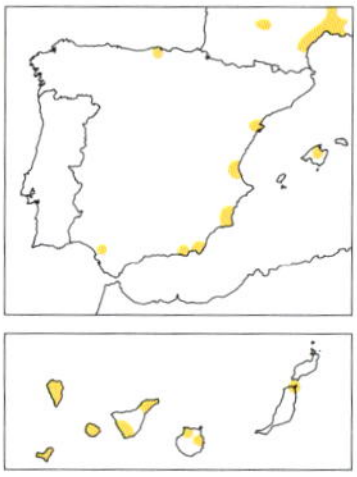

Common Tern *Sterna hirundo* (charrán común). ~35 cm. The only frequent red-legged tern in Spain. Considerably smaller than Sandwich Tern (above) and uncrested. Bill long and slender; red with black tip in summer, otherwise all dark. Tail streamers barely reach wing tips at rest. Outer primaries have dark trailing edge below and form dark wedge above in flight in summer, less obvious at other times. In winter plumage shows white forehead and dark carpal bar. Juvenile has orange bill base, brownish mantle and dark carpal bar ✦ Colonial nester on sandy coasts, saltpans, rocky islets in Canary Is. Fishes in bays, coastal lagoons, estuaries, ricefields ✦ In summer and on passage (Mar–May and Jul–Oct), wintering in W and S Africa ✦ Fairly common (some thousands of pairs, fluctuating) on Mediterranean coast (Delta, Albufera, Santa Pola); has bred Balearics; a few dozen pairs nest in Canary Is; nests occasionally in N and inland; **NT**. Common on coasts on passage, with rare winter records. Migrants rare inland and in the Balearic and Canary Is.

Roseate Tern *Sterna dougallii* (charrán rosado). ~35 cm. Rare. Adults have very long tail streamers, extending well beyond wingtips at rest; bill black, with red base in summer. Very pale above; breeding birds show pale pinkish wash below; primaries lack dark trailing edge below; 2–3 outermost primaries form slender dark wedge in autumn. Juvenile shows bold black feather edging on mantle and scapulars (as in Sandwich Tern), complete blackish cap until autumn, when forehead paler; carpal bar variably obvious; legs black ✦ Passage (Jul–Oct and Apr–May) of birds from small populations of British Isles and Brittany; usually well offshore ✦ Rare and irregular in autumn in coastal NW, esp. Asturias and Cape Bares, and SW; very rare in spring. Vagrant to Mediterranean and Canary Is. Nested formerly in El Hierro.

Arctic Tern *Sterna paradisaea* (charrán ártico). ~35 cm. Rare. Resembles Common Tern but smaller and more delicate; tail streamers project beyond wingtips in adults at rest; legs very short; bill all red when breeding. Greyish-tinted underparts contrast with whiter face in breeding season; upperwing lacks dark wedges; underwing very white, with neat black trailing edge to primaries; all primaries translucent. Juvenile greyish (not brownish above), bill all dark; carpal bar diffuse ✦ Offshore migrant ✦ On passage (Aug–Nov and Apr–Jun). Nests in N Europe, Siberia, Iceland, Greenland, etc., wintering around Antarctica ✦ A few are seen from Biscay coasts, usually from headlands, in autumn but probably very common offshore. Sporadic records elsewhere and at other times. Vagrant to Canary Is (but probably passes offshore).

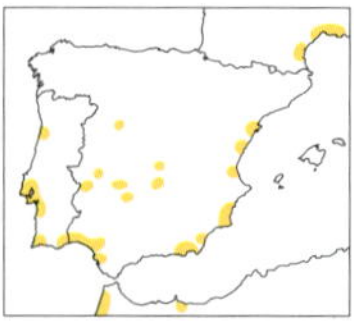

Little Tern *Sternula albifrons* (charrancito común). ~23 cm. Very small, bill and legs yellow, forehead always white. Outer primaries black. Flight erratic, with rapid wingbeats and frequent hovering. Marsh terns (below) are similar-sized but have smaller heads, shorter bills and less forked tails. Juvenile barred above with contrasting paler secondaries in flight. Noisy; calls sharp and strident: *kit, kit* or *kirri–kirri–kirri* ✦ Coastal or inland, with colonies at saltpans, littoral lagoons, quiet beaches, islands in reservoirs. Also forages in ricefields, marismas, estuaries ✦ In summer and on passage (Apr–May and Jul–Oct), wintering in W Africa ✦ A few thousand breed, fluctuating, quite concentrated (SW Andalucía, Albufera, Delta, SE wetlands); scarce nester inland (tens or a few hundred, most in Extremadura). Scarce in Balearics and Canary Is. **NT**

Common Tern
summer
juv.
winter
juv.
Roseate Tern
summer
winter
juv.
Arctic Tern
summer
winter
winter
Little Tern
summer
juv.

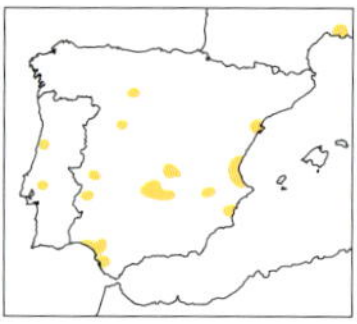

Whiskered Tern *Chlidonias hybrida* (fumarel cariblanco). ~25 cm. Most commonly encountered marsh tern and only one with a white 'face' and ash-grey underparts in breeding plumage. Wings short, tail very short. Bill and legs red, black in winter. Larger and longer billed than other marsh terns, with faster, more direct flight. Adult winter recalls Common Tern but has grey, slightly forked tail; paler than winter Black Tern, with little black on head and lacking carpal bar and dark breast-side patches. Juvenile shows brown mantle and scapulars, coarsely barred black; moulting this 'saddle' before migrating. Like other marsh terns, catches insects in flight over wetlands and picks food from surface, sometimes catching fish by shallow plunge-dives. Nests in noisy colonies, often alongside Black-necked Grebes ✦ Nests at shallow coastal and inland wetlands with floating vegetation. Visits ricefields ✦ In summer and on passage (Mar–May and Aug–Oct); a few winter. Most winter in W Africa ✦ A few thousand pairs breed, fluctuating greatly according to water levels and very localised (Marismas; to a lesser extent in the Delta, Valencia and La Mancha lagoons, irregularly in Extremadura and Castilla y León); **VU**. Common on passage on E coast and in the Balearics, more so in spring; scarce or rare in N. Up to a few hundred overwinter, most at Delta. Occasional on passage in Canary Is.

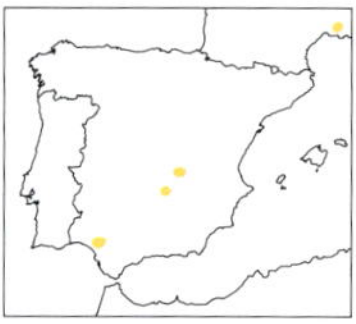

Black Tern *Chlidonias niger* (fumarel común). ~24 cm. Only common on passage. Breeding plumage very dark, especially in males; head and breast blackish, wings paler, undertail white; bill and legs black. Body plumage blotchy during moult (Jun–Aug). Adult winter has distinctive head and neck pattern; black cap and ear coverts contrasting with white forehead; has dark breast-side patches at wing bases. Juvenile similar, with dark grey, barred 'saddle' and prominent carpal bar. Flight buoyant. Rarely dives ✦ Nests in shallow, clean-water marismas, lakes and pools with abundant emergent vegetation. On passage also on ricefields, saltpans and seashores (estuaries, beaches, harbours) ✦ Chiefly on passage (esp. Apr–May and Aug–Oct), with a few summering and very rare winterers ✦ Formerly bred commonly but now nests only occasionally, irregularly and very locally (Marismas, La Mancha); **EN**. Still common or very common on passage on Mediterranean coasts and the Balearics, particularly in spring. Abundant in the Strait in autumn in some years but much rarer further west and in N. Irregular and scarce in Canary Is.

White-winged Black Tern *Chlidonias leucopterus* (fumarel aliblanco). ~22 cm. Quite rare. Breeding plumage striking, with inky black head, body and underwing coverts contrasting with white forewings, rump and tail. Bill shorter than in other marsh terns, dark red in summer. Legs red, paler in winter and juveniles. Autumn adult retains traces of breeding plumage including black underwing coverts and white rump; also paler-headed than autumn Black Tern and lacks breast-side patches. Juvenile has very dark saddle, a carpal bar and white rump ✦ Saltpans, estuaries, coastal lagoons ✦ Migrant (esp. Apr–May and Aug–Sept) with rare observations at other times ✦ Scarce but relatively regular on E and S coasts and in Balearics. Occasional inland. Vagrant in N. Vagrant to Canary Is.

winter
juv.
winter
Whiskered Tern
summer
Black Tern
winter
juv.
summer
winter
winter
White-winged Black Tern
summer

SKUAS (Charadriiformes, Stercorariidae) – Marine, gull-like predators and kleptoparasites (robbing food from other birds). Feet webbed; bill and claws hooked. Wings long and pointed. Ground-nesters in boreal zones. Long-distance migrants.

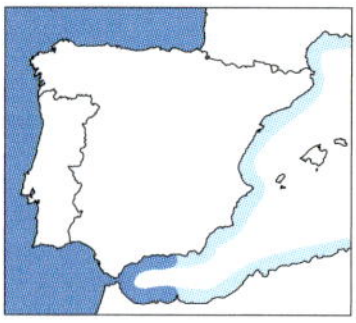

Arctic Jaeger/Skua *Stercorarius parasiticus* (págalo parásito). 41–45 cm. Common. Intermediate between Pomarine and Long-tailed Skuas in size, bulk, bill size and extent of white wing flashes. Adult has extended, pointed, central rectrices. Has dark (majority), intermediate and pale morphs. Juvenile has barred underwing coverts and axillaries; very similar to other juvenile skuas and best distinguished by structure. Rapid flight recalls falcons ✦ Marine, more coastal than other skuas ✦ On passage (Aug–Nov and Mar–May) and in winter; some over-summer. Nests north from Scotland and Scandinavia ✦ Common in autumn (Sept–Oct) on Biscay coast (Cape Peñas, Cape Bares) and in the Atlantic; Uncommon in the Mediterranean. Scarce winterers occur in Catalonia, Levant, Alborán, Strait. Scarcer in spring. Occasional inland and in Balearics. Scarce in Canary Is.

Pomarine Jaeger/Skua *Stercorarius pomarinus* (págalo pomarino). 45–50 cm. Sturdy, approaching Great Skua in size. Adults have elongated, twisted, spoon-shaped, central rectrices; less evident in winter. Deep-chested, wings broad-based with conspicuous white flashes; head large; bill heavy, bicolored. Has pale (some 90% of birds), intermediate and dark morphs; pale morph often shows marked breast-band. Juvenile heavily barred below on wings and body, also showing a wing flash on underwing primary coverts. Flight slower and steadier than in Arctic Skua ✦ Seldom inshore ✦ Migrant (Aug–Nov and Mar–May). Nests in Arctic ✦ Quite common in autumn off northern headlands, esp. during W winds; scarcer in the Mediterranean and Strait region. Scarce in spring, but more frequent then in Catalonia and Valencia. Vagrant to Balearic Is. Irregular and scarce in Canary Is.

Long-tailed Jaeger/Skua *Stercorarius longicaudus* (págalo rabero). ~50 cm. Rarest skua. Comparatively small, lightly-built, wings long and narrow, recalls terns. Adult has greatly elongated central rectrices and lacks obvious wing flashes. Pale morph (dark is very rare) has strongly contrasting dark cap, lacks breast-band; dark remiges contrast above with paler coverts. Juvenile grey-toned, with short tail projection; barred overall, especially strongly on undertail coverts ✦ Pelagic, hard to see from shore ✦ Autumn migrant (late Jul–Nov, most Aug–Sept). Nests in mountains of N Scandinavia and on shores of White Sea, Siberia, Greenland... ✦ Seldom recorded until recently (p. 233), from Biscay coasts and in the Mediterranean, but seems to be regular inshore off headlands in Asturias and Galicia. Has been found inland. Exceptional in spring (Apr–May). Vagrant to the Balearic and Canary Is.

Great Skua *Catharacta skua* (págalo grande). ~55 cm. Sturdy, brown, variably dark. Recalls a juvenile large gull but heavier in build, with shorter and broader wings, showing striking white flashes. Body barrel-shaped, thick-necked; tail short; bill stout and dark. 1st year resembles adult but unbarred, plain rufous below, with smaller wing flashes ✦ Generally offshore ✦ In winter and on passage (Aug–Dec and Feb–Apr), some over-summer. Breeds in Scotland, Iceland, Faroes, Norway... ✦ Quite common off all coasts in autumn and winter. Less common in spring, although more frequent then in the Mediterranean). Very rare inland. Very scarce in the Balearics. Regular, but scarce in Canary Is.

Arctic Jaeger/Skua
Pomarine Jaeger/Skua
♀
Long-tailed Jaeger/Skua
Great Skua

Skuas. Adult plumages

Skuas. Juvenile plumages

AUKS (Charadriiformes, Alcidae) – Fish-eating seabirds that 'fly' underwater with their wings. Resemble penguins in having rear-set legs, vertical stance and black-and-white plumage. Nest colonially on inslets and cliffs. Holarctic, frequently boreal.

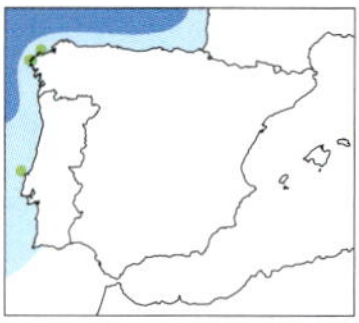

Common Murre/Guillemot *Uria aalge* (arao común). ~40 cm. Resembles Razorbill but slender, pointed bill evident at short range. Chocolate-brown above, with streaked flanks. In winter has a dark line crossing white ear coverts from eye. Bill plain black. Feet project beyond short tail in flight. Breeding plumage (acquired early in winter) has uniformly dark head and neck. 1st winters have smaller bills and less flank streaking. Flight rapid and direct, with whirring wingbeats, as other auks ✦ Nests on cliffs on coastal islets. Keeps well offshore in winter but over continental shelf ✦ Mainly in winter (Oct–Apr) and on passage; some over-summer. Arrive mainly from British Isles (ssps. *aalge* and *albionis*) ✦ Ssp. *albionis*: nested until very recently in Galicia (Cape Vilán, Sisargas Is) but recently extinct after prolonged decline; **CR**. In winter fairly common off Biscay coasts, scarce in Galicia and very scarce in Andalucía and the Mediterranean. Vagrant to Balearic and Canary Is.

Razorbill *Alca torda* (alca común). ~40 cm. Common in winter. Sturdier and larger-headed than Common Guillemot; bill deep and blunt, has a vertical white line. Bill smaller and all black in 1st winter. Unlike Guillemot is blacker; longer tail covers feet in flight and often raised when swimming; lacks flank streaking and has shorter auricular stripe in winter; axillaries mainly white; wingbeats even faster than Guillemot ✦ Continental shelf waters; often inshore in bays, estuaries, harbours ✦ In winter and on passage (Oct–Dec and Feb–Apr); most originate from British Isles (ssp. *islandica*) ✦ Very common on Biscay and Atlantic coasts and common in the Mediterranean, with winter concentrations off Asturias, Galicia, W Andalucía and Catalonia. Often washed up dead on beaches. Large-scale passage evident at Cape Peñas, Cape Bares, Strait, etc. Scarce in Balearics. Vagrant to Canary Is.

Atlantic Puffin *Fratercula arctica* (frailecillo atlántico). ~30 cm. Dusky while in Spanish waters. Some 25% smaller than Razorbill. Large head is all dark in winter. Short neck is always black. Broad-based, triangular bill is yellowish in winter adults, smaller and blackish in 1st winters. Underwings greyish; secondaries lack white border. Flies swiftly, often changing body angle ✦ Seldom inshore ✦ In winter and on passage (Oct–Dec and Mar–Apr); a few over-summer. Nests north from Brittany and British Isles (ssp. *grabae*) ✦ Uncommon in N and scarce in the Mediterranean; perhaps largely undetected on account of pelagic habits. Many thousands pass through the Strait in spring and autumn. Scarce in Balearics. Rare in Canary Is.

Little Auk *Alle alle* (mérgulo atlántico). ~20 cm. Occasional. Starling-size with tiny bill. Looks compact and neckless on the water. Tail small, pointed. In winter white of foreneck extends to ear coverts and shows a narrow black half-collar; scapulars have 3 or 4 white streaks ✦ Avoids warm waters; reaches coasts mainly when storm-driven ✦ In winter (Nov–Feb). Nearest major breeding colonies in Spitsbergen; most winter very far N ✦ Very scarce off Galicia and Biscay coasts; many records are of tideline corpses. Exceptional in Andalucía. Vagrant to the Mediterranean and Canary Is.

summer
Common Murre/Guillemot
winter
summer
winter
Razorbill
summer
Atlantic Puffin
winter
summer
Little Auk
winter

BARN OWLS (Strigiformes, Tytonidae) – Medium-sized nocturnal raptors. Bills relatively long and slender. Middle claw serrated. Chiefly hunt mice and other micro-mammals, expelling their bones and fur as pellets. Cosmopolitan.

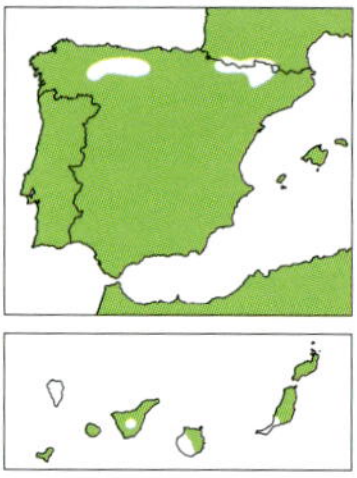

Common Barn-owl *Tyto alba* (lechuza común). ~35 cm. Familiar, 'white' owl. Facial disc heart-shaped, eyes black, wings and legs long. Underparts normally very white (ssp. *alba*) but yellowish in Lanzarote and Fuerteventura (*gracilirostris*) and ochre-brown and very spotted in birds of central European origin (*guttata*); intermediate colorations known from Vizcaya, Basque country. Produces eerie shrieks at night. Hunts low-down with slow, wavering flight. Often killed on roads. Pellets large (~4–6 cm × 2–2.5 cm), cylindrical, rounded at both ends, smooth, black and shiny when fresh ✦ Farmland, pastures; avoids closed woodland and mountains. Nests in belfries, lofts, barns, other farm buildings, escarpments, hollow trees ✦ Spanish birds sedentary; joined in winter, perhaps irregularly, by birds from central Europe (mainly *alba*) ✦ Ssp. *alba*: breeds commonly (tens of thousands of pairs, perhaps declining), widespread but most frequent in C and S. Common in Balearics; Uncommon in C and W Canary Is (very few records from La Gomera and La Palma). Ssp. *gracilirostris*: Uncommon; in Fuerteventura, Lanzarote and nearby islets, threatened; **EN**

OWLS (Strigiformes, Strigidae) – Comprise most nocturnal raptors. Bills hooked, all talons unserrated, eyes very large, plumages cryptic. Size, and prey types, very variable; small mammals predominate in diet. Produce pellets. Nidicolous. Worldwide.

Little Owl *Athene noctua* (mochuelo europeo). ~23 cm. Very common and often seen in daytime. Compact, 'earless', with large yellow eyes and fierce expression; white spotted; long-legged. Juvenile paler and more uniformly coloured. Perches on fence posts, stone heaps, hut roofs. Crouches and nods briskly when alarmed. Noisy in winter and early spring, making mewing calls at dusk, including a plaintive *kíu*. Flight low and undulating, closing wings between wingbeat bouts. Pellets 2–4 cm × 1–1.5 cm, grey when fresh. Often killed on roads ✦ Habitat varied: selects open areas (croplands, pastures, fallows) offering nest sites (holes in trees, scree slopes, stone walls, rabbit burrows); avoids forests and mountains ✦ Sedentary; with short-range juvenile dispersal ✦ Ssp. *vidalii*: widespread and very common (tens of thousands of pairs, declining); most numerous in warm, even semi-arid, areas. Scarce in Mallorca and Menorca. Ssp. *glaux*: nests in Ceuta and Melilla. Doubtfully recorded in Canary Is.

Boreal Owl *Aegolius funereus* (mochuelo boreal). ~25 cm. Pyrenees. Resembles Little Owl but habitats and range are distinct. Large-headed, facial disc has dark margin and extends well above eyes. Juvenile very uniform chocolate-brown. Entirely nocturnal, roosting hidden in foliage. Calls at night (esp. Jan–Apr): a far-reaching *pu–pu–pu–pu–pu* (3–6 notes) repeated every few seconds ✦ Subalpine coniferous forests, chiefly Black Pine, at altitudes of 1,700–2,200 m (also in deciduous and mixed woods above 800m in Arán valley). Nests in tree holes (often old Black Woodpecker nests) and nestboxes ✦ Sedentary, with some dispersion ✦ Rare (a couple of hundred territorial males), chiefly in Catalan Pyrenees (Lleida province), where first discovered in 1980s.

Common Barn-owl

Little Owl

Boreal Owl

Eurasian Scops-owl *Otus scops* (autillo europeo). ~20 cm. Smallest Spanish owl, ~10% smaller than Little Owl. Sings in spring, occasionally even by day; a short, clear whistle, *tiu* or *kiu*, repeated every 2–3 s for long periods (can be confused with similar call of Midwife Toads, *Alytes* spp., although this is briefer and more metallic). More slender than Little Owl and has ear tufts; lacks 'frowning' face; streaked below. Ground colour ranges from grey to rufous. Hard to see. Small pellets (1–1.5 cm × 0.5–1 cm) contain chitinous insect remains ✦ Riparian woodlands, copses, dehesas, orchards, parks, etc., generally at low elevations; avoids dense forests. Nests in tree holes, nestboxes, corvid nests and occasionally holes in walls ✦ Summer visitor (Mar/Apr–Aug/Oct) with some over-winterers and perhaps some passage of central European birds (ssp. *scops*) ✦ Ssp. *mallorcae*: breeds commonly (a few tens of thousands of pairs, declining), most frequent in coastal regions and scarce or rare in N and montane regions. Common in Balearics; present in Ceuta and Melilla. Some winter, chiefly on Mediterranean coasts, the Balearics and locally in Andalucía and Extremadura. Occasional on passage in Canary Is, most often in eastern islands.

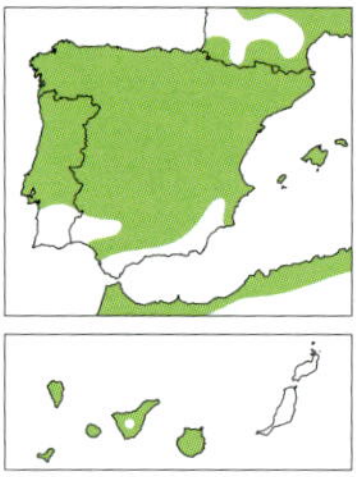

Long-eared Owl *Asio otus* (búho chico). ~35 cm. Has ear-tufts, like Eagle Owl, but much smaller. Slimmer than Tawny Owl, with orange eyes and rusty-brown facial disc. Nocturnal song, seldom heard but most often in winter: a low *huu* every 2–3 s. Hides in tree cover by day. Usually solitary but forms communal roosts in winter. Pellets similar to those of Tawny Owl but somewhat smaller (~4–5 cm × 2 cm) and more irregular ✦ Isolated pine copses, plantations, scrub-oak stands, riverine woodlands, woods with clearings; often near croplands. Uses old corvid nests in trees ✦ Spanish birds sedentary. Some winterers, probably not many, from C and N Europe ✦ Ssp. *otus*: uncommon (a few thousand pairs); probably widespread but easily overlooked; seems commoner in the Duero and Ebro valleys. In Balearics is common in Mallorca, Ibiza and Formentera. Ssp. *canariensis*: common and widespread in central and western Canary Is; **DD**

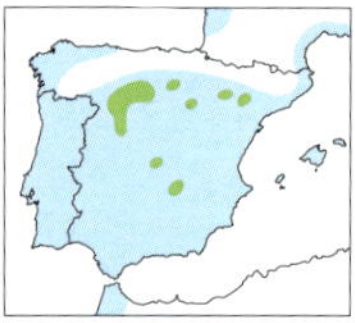

Short-eared Owl *Asio flammeus* (búho campestre). ~37 cm. Slightly smaller than Long-eared Owl with barely noticeable ear-tufts; Eyes yellow, bordered black (esp. in juveniles); strongly streaked breast contrasts sharply with very pale belly. In flight wings look longer and narrower than in Long-eared Owl, with fewer but broader bars on remiges giving blackish wingtip effect; often raised in a V; tail more boldly barred. Hunts over open country, often by day, at times in groups. Nuptial song, often given in display flight: a gentle *buu–buu–buu–buu*… ✦ Nests on ground, in wetlands, sedgy meadows, fallows ✦ Wintering and passage (Sept–Nov and Mar–Apr) of W European birds; with over-summerers and some breeders ✦ Scarce in winter, most often in northern half and at wetlands. Passage inconspicuous. Only nested sporadically until 1994, a Common Vole (*Microtus arvalis*) plague year, when some hundreds of pairs nested; has bred regularly since then but in very variable numbers; highest breeding density in northern Meseta; **NT**. Scarce or rare in Balearics (has bred). Irregular and scarce winterer in Canary Is.

Eurasian Scops-owl
grey morph
brown morph
Long-eared Owl
Short-eared Owl

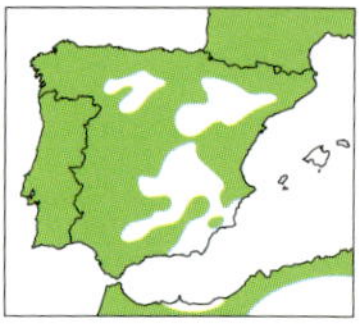

Tawny Owl *Strix aluco* (cárabo común). ~40 cm. Common in woodlands. The familiar wavering hoot of the male: *huu*, *hu*, *huhuhuhuuu* is heard year-round, but particularly Feb–Apr; contact call: *ke-wick*. Medium-sized, compact, eyes dark, lacks ear-tufts. Wings shorter and broader than in Long-eared and Short-eared Owls; remiges barred below. Two colour morphs: grey and rufous. Pellets ~5–6 cm × 2 cm, grey, lumpy ✦ Any mature woodlands, coniferous or deciduous, at all elevations. Also open countryside with trees, riparian woodlands, parks. Nests in tree holes, nestboxes, sometimes in holes in buildings or cliffs ✦ Sedentary, with short-distance juvenile dispersion ✦ Ssp. *sylvatica*: common (some tens of thousands of pairs, perhaps increasing); much commoner in N and montane areas. Rare in treeless regions. Absent from Balearic and Canary Is. Ssp. *mauritanica*: nests in Ceuta and Melilla.

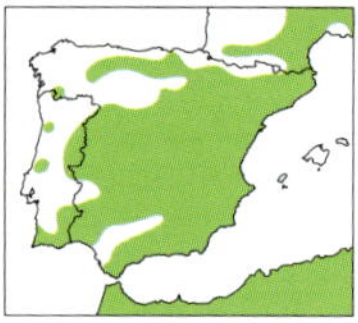

Eurasian Eagle-owl *Bubo bubo* (búho real). 60–70 cm. Only very large owl in Spain. Ear-tufts long; eyes orange, large; talons powerful. Well known song: a loud, far-reaching, deep *u-hu* repeated every 8–10s, given most often at sunset and early in the year; food-begging call of young also far-reaching: an incessant *chiu* or *chiuyu*. Flies silently on broad, rounded wings, with rapid, deep wingbeats ✦ Hilly rocky areas, with cliffs or caves for nesting; sometimes nests on ground or buildings; seldom in high mountains or rainy areas lacking rabbits ✦ Sedentary but dispersive ✦ Ssp. *hispanus*: quite common (a few thousand pairs, increasing) but unevenly spread, most frequent in C, S and Mediterranean coastlands; rare or absent from large parts of N. Absent from Balearic and Canary Is.

OSPREY (Accipitriformes, Pandionidae) – Family comprises one cosmopolitan species. Foot adaptations for fish capture include reversible outer digits and spiny soles. Sexes similar, males smaller.

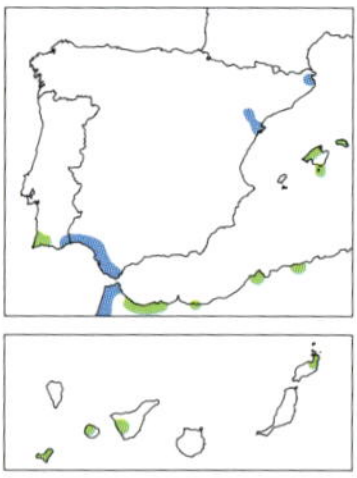

Osprey *Pandion haliaetus* (águila pescadora). 50–55 cm. Unusual raptor. Sizable but not much larger than a Yellow-legged Gull. Very white below, with long, narrow, distinctively angled wings, showing marked black carpal patches and dark greater coverts below. Head white with dark mask; variably streaked breast band, less marked in adult male; tail barred. Upperparts uniformly dark brown; with pale feather tips in juveniles. Hovers before plunging for fish feet-first from a height. Carries fish in talons, then often chased by gulls ✦ Native birds usually nest on coastal islets, on cliffs and stacks; introduced birds nest on trees or artificial platforms. Visits estuaries, reservoirs, gravelpits, saltpans, bays, rivers, etc. ✦ Adult native birds mainly sedentary but young dispersive. Passage (mainly Mar/Apr and Aug/Oct) and some wintering of birds from Sweden, Germany, etc. ✦ Scarce breeding (a few tens of individuals) in Balearics and Canary Is, irregularly in Chafarinas Is; **CR**. Formerly nested on coastal cliffs in E and S of Peninsula. Now reintroduced at Marismas (Odiel) and reservoirs of SW Andalucía, where first nested in 2009, increasing steadily. Broad-front migrant, most often noted in E and Strait region. Scarce but regular in winter (a few hundred individuals, increasing) in Huelva, Cádiz and the Mediterranean coastlands.

Tawny Owl

Eurasian Eagle-owl

Osprey

DIURNAL RAPTORS (Accipitriformes, Accipitridae) – Medium-sized, large or very large. Predators or carrion-eaters, with talons and hooked beaks. Wings often broad, with emarginated primaries, facilitating soaring flight. Variably migratory. Sexes alike but females larger. Cosmopolitan.

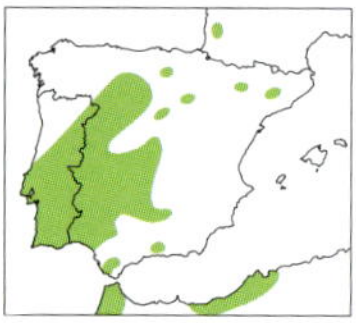

Black-winged Kite *Elanus caeruleus* (elanio común). 31–35 cm. Looks unusual, resembling a large-headed white pigeon; tail short; wings broad but pointed. Eye patch and areas of wings black. Juvenile less clean-cut, with white tips to dorsal feathers and rufous-tinged breast. Often seen perched on posts or hovering, with wings raised in a V. Solitary or in pairs, but forms communal roosts in winter ✦ Open holm oak dehesas with cereal crops, or croplands and pastures with scattered trees; often in irrigated zones in winter. Hunting often crepuscular. Nests in trees ✦ Sedentary but disperses in autumn–winter especially ✦ Uncommon (some hundreds of pairs, increasing). Population centred on Extremadura and nearby area but expanding (Northern Meseta, Andalucía, Aragón, Catalonia). Extremely rare in Canary Is. **NT**

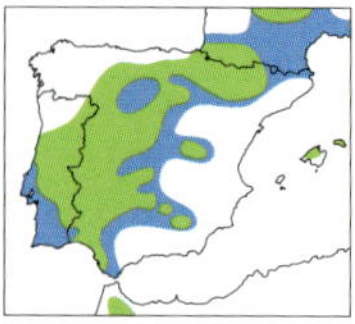

Red Kite *Milvus milvus* (milano real). 60–70 cm. Slim; wings and tail longer than Black Kite; tail deeply forked. Shows contrasting colours; large white panel at primary bases; tail reddish above, belly rufous, head grey. Juvenile shows pale feather edgings above; undertail coverts pale. Spends long periods in flight, patrolling slowly, characteristically twisting tail. Often in groups (at nesting and feeding areas; winter roosts). Long whistling calls striking ✦ Level or hilly country, with pines, dehesas, riparian woods or forest edges. Nests in tall trees. More widespread in winter, when often at rubbish tips and town outskirts ✦ Year-round. Spanish birds partially migratory; large numbers of winterers from Germany and nearby countries (Sept/Oct–Mar). Very few now cross the Strait ✦ Breeding population (a few thousand pairs) recovering after rapid decline attributed to illegal poison baits; concentrated in Extremadura, SW Castilla y León, Navarra/Huesca; absent from very humid or very arid regions. Very small numbers (a few tens) nest in Mallorca and Menorca, where slowly increasing. Extinct in Canary Is since 1960s; now vagrant there. Winter population of a few tens of thousands, declining; most in open country and farmlands of Extremadura and the Ebro and Duero valleys. **EN**

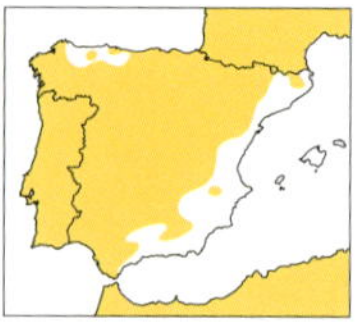

Black Kite *Milvus migrans* (milano negro). 50–60 cm. Most abundant kite in spring and summer. Less elegant than Red Kite and typically dark brown; confusable at long range with dark phase Booted Eagle (p. 130) or even Marsh Harrier (p. 132) but has forked tail; less obvious when spread. Juveniles pale-streaked below with white-tipped greater coverts above. Gives long, sharp trilling calls ✦ Countryside with some trees (riparian woodlands, dehesas), chiefly in lowlands and often near water. Nests in trees, more rarely on riverine cliffs, often in loose colonies. Forages opportunistically along rivers, at reservoirs, rubbish dumps, over farmland and along roads… Gregarious, forming large roosts ✦ Chiefly trans-Saharan migrant (passage mainly late Feb–May and Jul–Sept). Many tens of thousands cross the Strait, including populations of W Europe ✦ Breeds commonly (>15.000 pp., increasing; with numerous immature non-breeders). Abundant in W (Duero basin, Extremadura and W Andalucía) and Ebro valley; appears in Mediterranean coastlands largely on passage. Small numbers winter in S. Scarce but regular in Balearic and Canary Is. **NT**

Black-winged Kite
juv.
Red Kite
Black Kite

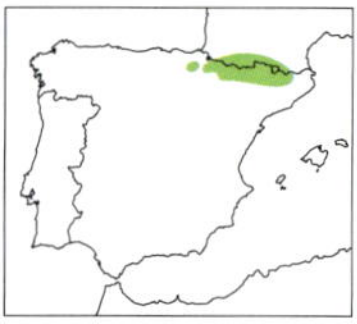

Bearded Vulture *Gypaetus barbatus* (quebrantahuesos). 100–120 cm. Imposing. Wingspan similar to Griffon Vulture but wings narrow and pointed; head prominent; tail long, wedge-shaped. Wings often pressed downwards when gliding, head and tail drooped. Juvenile and immatures darker and dark-hooded. Generally solitary. Flight elegant, often along mountain slopes ✦ Montane. Nests in caves on cliffs, in high-level, broken terrain. Searches for carrion and consumes bones; smashing large ones by dropping them on rocks ✦ Sedentary but immatures may wander ✦ Over one hundred pairs, increasing. Formerly found in many montane regions but now restricted to Pyrenees (mainly Huesca and Lleida) and Basque mountains; reintroduction in progress at Cazorla/Segura and Picos de Europa. **EN**

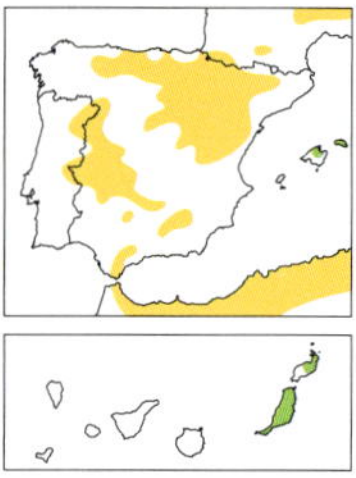

Egyptian Vulture *Neophron percnopterus* (alimoche común). 60–65 cm. Smallest vulture. Wings broad; overall white with black remiges; projecting yellowish head and neck; tail wedge-shaped (see similarly patterned White Stork and pale morph Booted Eagle). Juvenile brown, very dark below; immatures show transitional plumages, for up to 4 years. Agile in flight. Often in company with other vultures, kites and corvids. Roosts communally ✦ Mountains and valleys with escarpments; nests in small caves. Often around human habitation. Carrion-eater and opportunistic, visits rubbish tips and carrion dumps ✦ Mainly summer visitor (passage ~Mar–May and Aug–Sept). Island birds sedentary ✦ Ssp. *percnopterus*: over 1,000 pp.; regional trends differ (highly prone to poisoning); most abundant in N (Aragón, Navarra, N Burgos, Sistema Ibérico) but also numerous in Salamanca, Cáceres and Cádiz; Balearic populations small (mainly Menorca); **EN**. Ssp. *majorensis*: Canary Is (a couple of hundred birds, most in Fuerteventura); **CR**

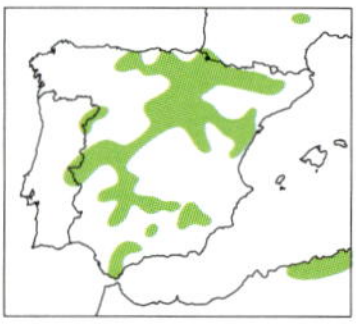

Griffon Vulture *Gyps fulvus* (buitre leonado). 95–105 cm. Very large. Glides on long, broad wings slightly raised in shallow V; bulging secondaries and long primary emarginations. Coverts paler than remiges, unlike in Cinereous Vulture. Smaller-headed and shorter-tailed than eagles. Juvenile somewhat darker, with bill grey; ruff brownish, not white and fluffy as adult. The African **Rüppell's Vulture** *Gyps rueppellii* (buitre moteado) appears occasionally, mainly in S ✦ Highly sociable. Nests in colonies, sometimes of dozens of pairs, on cliffs in mountains and gorges. Ranges widely in search of carrion ✦ Adults largely sedentary; ranging higher in mountains in summer; thousands, mainly immatures, migrate across the Strait (~Oct–Nov and Apr-Jun) ✦ Census (2008): ~25.000 pp., increasing. Common and widespread, except in Galicia and in lowland or littoral areas; abundant in Aragón, Navarra, E and S Meseta, northern half of Extremadura and parts of Andalucía. Has recently colonised Mallorca.

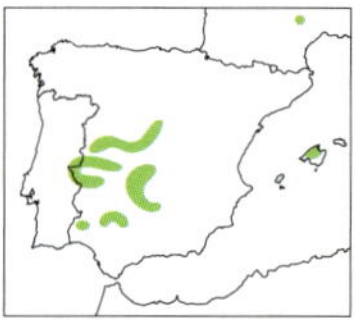

Cinereous (Black) Vulture *Aegypius monachus* (buitre negro). 100–110 cm. Majestic. Even larger than Griffon Vulture. Distinguished by very dark, largely uniform plumage; more rectangular wings and tail sometimes slightly wedge-shaped; underwing coverts blacker than remiges; glides on horizontal or slightly drooping wings. Juvenile blacker than adult; head and ruff blackish and bare skin areas pinkish, not greyish. Often accompanies Griffons, which generally outnumber them ✦ Wooded mountainsides, nesting in loose colonies; large nests built on trees (cork or holm oaks, pines) ✦ Adults sedentary; immature birds dispersive ✦ Locally common, 2,000+ pp., increasing; concentrated in Extremadura and nearby areas. Small population in Mallorca (100+ birds, increasing); **VU**

Bearded Vulture
juv.
ad.
ad.
Egyptian Vulture
2nd-3rd year
juv.
ad.
juv.
Griffon Vulture
ad.
juv.
Cinereous (Black) Vulture

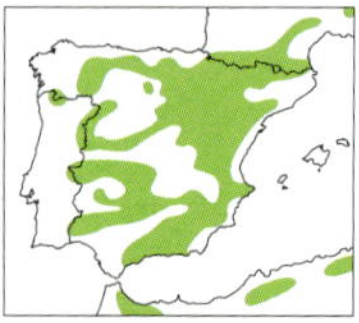

Golden Eagle *Aquila chrysaetos* (águila real). 80–90 cm. Very large and dark. Well proportioned, powerful; head and neck project prominently in flight; tail long, rounded. Adults nearly uniform brown, wih nape and rear neck yellowish or rufous brown. Juvenile unmistakable; has large white basal patches to primaries and white tail base; these reduced gradually over 5–6 years in subadult plumages. Soars and glides with slightly raised wings ✦ Mountains and broken country at very variable altitudes, including human-uninhabited lowlands. Nests on cliffs (~90%) or trees ✦ Sedentary with juvenile dispersion ✦ Ssp. *homeyeri*: quite common (>1.500 pp., increasing gradually), frequent in Pyrenees, Ebro valley; Cantabrian, Iberian and Central ranges, Sierra Morena and Betic mountains. Vagrant to Balearic Is. Possible vagrant to Canary Is. **NT**

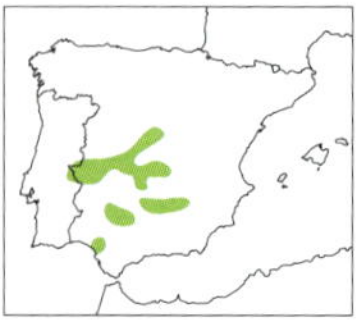

Spanish Imperial Eagle *Aquila adalberti* (águila imperial ibérica). 75–80 cm. Somewhat smaller and even darker than Golden Eagle; wings long, parallel-edged, tail long, quite straight-ended. Adult shows distinctive white 'shoulder' patches at close range. Juvenile very different; showing contrast between reddish-brown body and dark flight feathers (recalling Griffon Vulture but has pale inner primary wedge and white trailing edge to wings and tail). Adult plumage acquired via several immature plumages. Voice corvine: *goc… goc… goc…*, heard especially Jan–Mar ✦ Mediterranean woodlands (holm and cork oaks, pines), often in hilly terrain. Nests in trees ✦ Sedentary but dispersive ✦ Scarce (~500 pp.) and local (with significant concentrations in Madrid, Cáceres, Toledo mountains, Sierra Morena, Doñana), but now increasing and spreading. **EN**

Greater Spotted Eagle *Clanga clanga* (águila moteada). 65–70 cm. Rare (p. 234). Smaller than Golden and Imperial Eagles but sturdy and compact. Wings very broad, secondaries bulging, remiges blunt-ended, shows 7 'fingers' in flight. Tail shorter than wing breadth, slightly wedge-shaped. Glides with arched wings. Adult uniformly dark, with pale crescent at primary bases. Juvenile and immature show lines of white spots on wing coverts. Easily confused with **Lesser Spotted Eagle** *Clanga pomarina* (águila pomerana), a very scarce migrant (p. 234) which is less sturdy, has body paler than remiges and shows 6 'fingers' ✦ Near water (marismas, lagoons) in areas with trees ✦ In winter or on passage (~Sept–Apr) ✦ Very few records but a few individuals now annual in winter at E coast wetlands (El Hondo) and sometimes the Marismas.

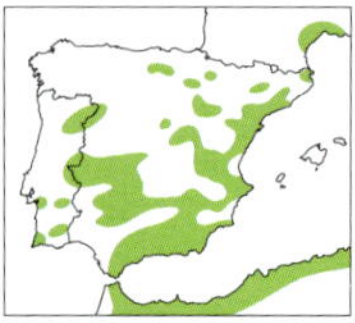

Bonelli's Eagle *Aquila fasciata* (águila perdicera). 60–70 cm. Large and powerful; wings shorter and broader than other eagles. White body contrasts with mainly dark wings. White mantle patch. When overhead, adult shows white leading edge to wings, variably pale lesser coverts; median and greater coverts forming a blackish band; remiges greyish, tipped darker; broad black terminal tail band. Juvenile very different: body and underwing coverts buff or rufous; primary coverts dark-tipped; tail densely and evenly barred. Transitional plumages occur for up 5 years. Solitary or in pairs; aggressive towards vultures and other birds ✦ Broken country at low or moderate elevations, with cliffs or gorges; occasionally nests in trees. Chiefly in warm, dry regions ✦ Sedentary, with juvenile dispersion ✦ Scarce (~750 pp.). Quite common along the Mediterranean littoral and in Castilla-La Mancha, Andalucía and Extremadura; rare and declining elsewhere; **EN**. Reintroduction underway in Balearics. Vagrant to Canary Is.

Golden Eagle
Spanish Imperial Eagle
Greater Spotted Eagle
Bonelli's Eagle

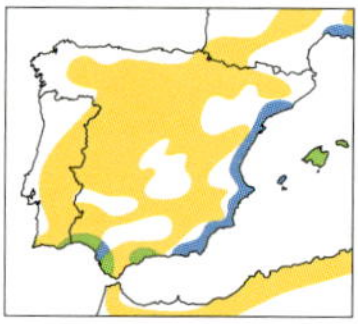

Booted Eagle *Hieraaetus pennatus* (águila calzada). 45–50 cm. Buzzard-size but eagle-like in proportions and actions. Two morphs, with base colour white or variably dark brown. Both show contrasting pale wedge on inner primaries, pale median covert band above, white uppertail coverts and white spots at front of wing bases. Light morph (~80%) unmistakable (but see Egyptian Vulture, p. 126). Dark morph records Black Kite (p. 124) and Marsh Harrier (p. 132) but tail paler, straight-ended; wings shorter; flight more deliberate ✦ Pine woods, holm oak woods, dehesas, riparian woodlands, etc. in areas of Mediterranean climate, most commonly in hilly regions. Nests in trees; on cliffs in Balearics ✦ In summer and on passage (Mar/May–late Aug/Oct); some overwinter. Sedentary in Balearics ✦ Common (thousands of pairs, increasing), most abundant in SW quadrant (SW Castilla y León, Central range, Extremadura, Andalucía). Hundreds of French-breeders occur on passage. Concentrates at Strait on migration. Small numbers overwinter, mainly around wetlands in E and SW. Small population (a few hundred birds) in Mallorca and Menorca. Scarce on passage and in winter in Canary Is.

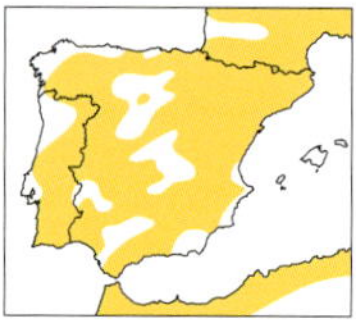

Short-toed Eagle *Circaetus gallicus* (culebrera europea). 62–67 cm. Large, very pale, often hovers. Long broad wings lack dark carpal patches, white or with rows of dark spots below, primary tips barely darker. Head broad; white in subadults, adults have dark brown hood, juveniles may have a rufous hood. Tail white, has 3-4 dark bars. Upperparts more uniform; brownish-grey. Calls whistling ✦ Hunts snakes on hillsides with low scrub, fallows and clearings. Selects hilly areas, in regions of Mediterranean or sub-Mediterranean climate ✦ In summer and on passage (Feb–May and Aug–Oct); a few overwinter ✦ Quite common (a few thousand pairs, increasing). Common in Cantabrian mountains, Pyrenees and sierras of Catalonia and Valencia, and in SW quadrant; scarcer in Biscay coastlands, mesetas, Guadalquivir valley and SE. A few winter in warmer littoral regions. Hundreds of French-breeders occur on passage. Concentrates at Strait on migration. Scarce migrant in Balearics. Vagrant to Canary Is.

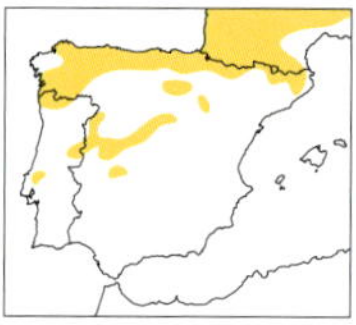

European Honey-buzzard *Pernis apivorus* (abejero europeo). 52–60 cm. Buzzard-like but slimmer; head narrower, cuckoo-like, projecting; tail longer. Plumage colour very variable, ranging from largely white to chocolate brown below. Below typically shows pronounced dark carpal patch, three or four bars across remiges and three tail bars. In adults, heads grey in males, brown in females. Most juveniles show dark secondaries; wings evenly barred below; slimmer-winged and shorter tailed than adults. Strong flier; soars on flat wings, broad arm wider than narrower hand; glides on slightly arched wings. Migrates in flocks. Discrete on breeding grounds. Often flushed from ground, where digs out bee and wasp nests ✦ Woodlands (beech, oak, pine) interspersed with pastures and croplands ✦ In summer and on passage (mainly late Apr–May and Aug–Sept) ✦ Uncommon breeder (some hundreds of pairs), most abundant in Pyrenees, Cantabrian range and Galicia. W European migrants follow the Pyrenean passes and the Mediterranean coast, crossing the Strait, where abundant, and Alborán basin. Scarce but regular migrant in Balearics. Vagrant to Canary Is.

Booted Eagle
Short-toed Eagle
European
Honey-buzzard
pale morph

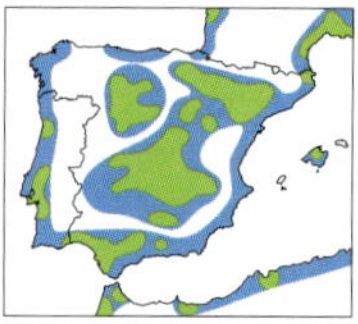

Western Marsh-harrier *Circus aeruginosus* (aguilucho lagunero occidental). 45–55 cm. Larger and less agile than other harriers; also shares long tail and long, narrow, parallel-edged wings, held in shallow V when soaring or gliding. Adult male tricoloured (brown, grey, black). Female and juvenile chocolate brown; crown and throat creamy. Females show white on forewings, recalling Imperial Eagle. Juvenile has very dark underwing coverts. Female/juvenile confusable with Black Kite, which has forked tail (p. 124) and dark morph Booted Eagle, which is more compact (p. 130) ✦ Nests in reed and reedmace beds in marshes, lakes, reservoirs, rivers, etc., and sometimes in cereal fields. Visits ricefields, pastures, fields ✦ Spanish birds resident. Birds from C & N Europe occur on passage and in winter (Sept/Oct–Mar/Apr) ✦ Breeding birds fairly scarce (~a thousand pairs, increasing) but widespread and now quite common in Ebro valley, the Meseta and W Andalucía; rare breeder on Mediterranean coast and in the Balearics (Mallorca); nests locally elsewhere. Commoner in winter (some thousands), chiefly in E and S (Delta, Albufera, Marismas, La Mancha). Rare but regular in winter and on passage in Canary Is.

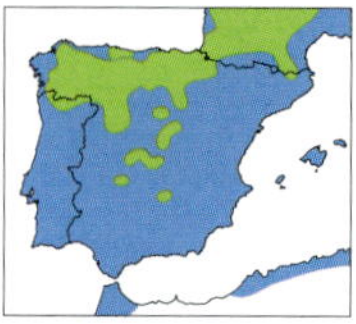

Hen Harrier *Circus cyaneus* (aguilucho pálido). 45–50 cm. Larger than Montagu's Harrier with broader, blunter wings; shows 5 'fingers' in flight. Adult male unmistakable; pale grey above; unstreaked white below, primaries extensively black, wing trailing edge dark grey below; rump white. Female brown with large white rump. Juvenile resembles female; ochre-coloured below, streaked dark ✦ Nests on ground, chiefly in scrub (heathland, gorse) or young conifers, often at low and medium elevations; in cereal crops in some areas. Widespread in open country in winter, roosting communally at wetlands ✦ Year-round, with wintering European birds and inconspicuous passage (Sept–Oct and Mar–Apr) ✦ Uncommon (~a thousand pairs), most frequent in N Navarra, Basque Country and N Castilla y León; isolated pockets of breeders on S Meseta and in Extremadura. Quite common and widepread in winter (a few thousand). Scarce visitor to Balearics. Rare and irregular in Canary Is.

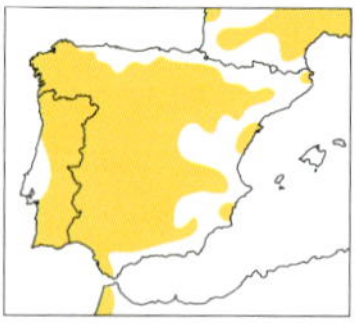

Montagu's Harrier *Circus pygargus* (aguilucho cenizo). 40–45 cm. Flight very buoyant on narrow wings; shows 4 'fingers'. Male has two black bars on secondaries below, one bar above; also rufous-streaked below and lacks white rump, unlike male Hen Harrier. Female more similar to female Hen but note wing formula; also lighter in build, with obvious facial pattern; dark ear coverts and white eye surround; white rump less obvious. Juvenile distinct; unstreaked rusty below. Second-year males show a transitional plumage. Has blackish-grey dark morph, most frequent in NW. Similar **Pallid Harrier** *Circus macrourus* (aguilucho papialbo), rare but increasingly frequent on passage and in winter, is a confusion species (p. 234) ✦ Cereal croplands especially, sometimes extensive pastures; also heathland, gorse scrub, young pine scrub, salt flats, etc., chiefly at low elevations. Nests on ground ✦ In summer and on passage (Mar/Apr–Aug/Sept) ✦ Common (some thousands of pairs), now probably increasing. Most numerous in W half (Extremadura, Andalucía, parts of Castilla, inland Galicia) but rare in montane areas or littoral zones. Scarce migrant in Balearics. Irregular and scarce in Canary Is, on passage and in winter. **VU**

Western Marsh-harrier
♂
♀
Hen Harrier
♂
♀
Montagu's Harrier
♂
♀

Northern Goshawk *Accipiter gentilis* (azor común). 50–55 cm. Combines long, rounded tail with broad wings; narrower at hand. Female as large as Common Buzzard, but male much smaller. Deep-chested and long-necked; dark-capped appearance above strong white supercilium. Juvenile buffish below, boldly streaked brown; lacks dark carpal patches of Common Buzzard and Honey-buzzard. Secretive, except during nuptial flight (Feb–Mar), when white undertail coverts striking. Climbs with frequent wingbeats ✦ Pine, beech and oak woods, including holm oaks, preferably with clearings, at any level ✦ Sedentary but dispersive ✦ Uncommon (a few thousand pairs), most abundant in N and in montane regions.

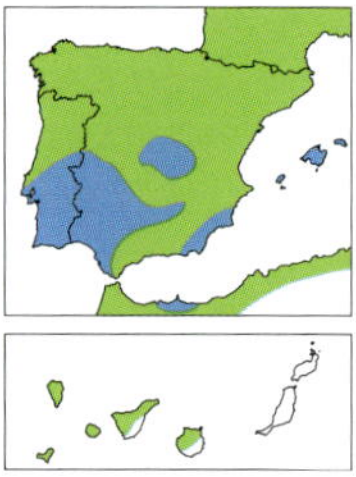

Eurasian Sparrowhawk *Accipiter nisus* (gavilán común). 30–35 cm. Kestrel-sized, with short, rounded wings. Tail long, square-ended. Males up to 25% smaller than females; barred rufous below. Juvenile similar to female but browner; more coarsely barred below. Flies low, bouts of rapid wingbeats alternating with short glides; climbs rapidly in tight circles ✦ Woods (pines, oaks) and isolated copses, often very small ones or young plantations ✦ Spanish birds resident. Some numbers of European birds occur on passage and in winter (Sept/Nov–Mar/Apr) ✦ Ssp. *nisus*: nests commonly (some thousands of pairs, increasing) in mountains of N and E but scarcer elsewhere and absent from extensively cultivated plains. Commoner and more widespread in winter. Ssp. *granti*: central and western Canary Is; uncommon, in broadleaf and pine woods, sometimes in farmland; **VU**

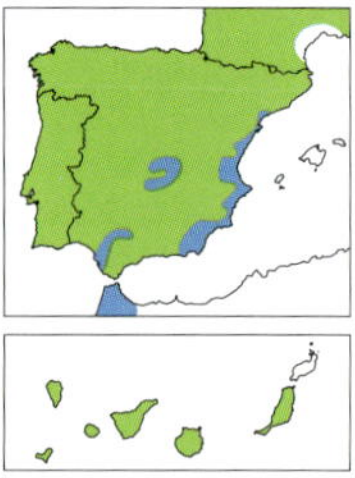

Common Buzzard *Buteo buteo* (busardo ratonero). 50–55 cm. Very widespread. Compact, wings broad and rounded, head broad but not prominent; tail relatively short. Brownish, variably dark or pale. Carpal patches dark, flight feathers show pale bases, pale breast band typical. Adult shows dark trailing edge to wings and sometimes dark tip to barred tail; these lacking in juvenile. Often perches at roadsides. Soars with spread tail and wings raised in shallow V; sometimes hovers. Calls mewing ✦ Woods, dehesas, open terrain with trees; in winter, often in treeless open farmland ✦ Spanish birds resident with some seasonal movements. Passage and wintering (Sept/Oct–Mar/Apr) of N European birds ✦ Ssp. *buteo*: common breeder (over 20,000 pp.); widespread, especially in N and W; except in treeless regions. Ssp. *vulpinus*: Occasional or rare migrant. Ssp. *insularum*: Canary Is. Common in woods and escarpments (several hundred pairs); **NT**

Long-legged Buzzard *Buteo rufinus* (busardo moro). 43–50 cm. Resembles Common Buzzard, especially *B. b. vulpinus,* which is very rare in Spain. Very pale head and breast contrast with rufous belly and flanks; tail unbarred, rufous. Carpal patches dark. Wings show clear dark trailing edges. Bill strong and legs relatively long. Spends long periods on posts and holds wings in V when gliding, like Common Buzzard; also hovers ✦ Steppes and all woodland types ✦ Resident but dispersive, with some nomadism and migration ✦ Ssp. *cirtensis* (endemic to NW Africa): widespread but uncommon in Morocco. Scarce resident in Ceuta. Vagrant to the Peninsula (p. 234), except extreme south where scarce but regular; has recently nested near the Strait.

Northern Goshawk
♀
juv.
Eurasian Sparrowhawk
♀
juv.
♂
Common Buzzard
Long-legged Buzzard

Raptors in flight

subad.
ad.
juv.
Spanish Imperial Eagle
subad.
ad.
Golden Eagle
ad.
subad.
1st winter

Black Kite
ad.
juv.
Red Kite
juv.
ad.
Booted Eagle
ad.
ad. dark morph
ad. pale morph

Hen Harrier
juv.
ad. ♀
ad. ♂
juv.
Montagu's Harrier
ad. ♀
juv.
ad. ♂
ad. ♂
juv.
ad. ♀
Western Marsh-harrier

HOOPOES (Bucerotiformes, Upupidae) – Two species; one widespread in Eurasia and Africa, the other only in Madagascar. Related to hornbills and woodhoopoes. Terrestrial insectivores.

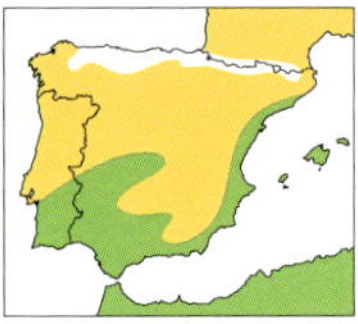

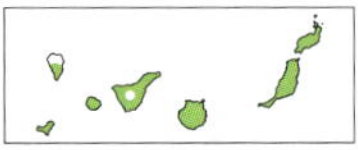

Common Hoopoe *Upupa epops* (abubilla común). ~27 cm. Unmistakable. Flight recalls large butterfly, on broad, rounded, black and white wings; displays erectile crest. Song distinctive: *hu–pu–pu*, low-pitched but far-reaching; repeated at short intervals. Bill long and decurved; shorter in juveniles. Females often browner below and white-throated ✦ Open woodlands, dehesas, pastures, croplands. Nests in holes in trees, stone walls and ruins ✦ Migratory, most wintering in Africa (passage Feb–Apr and Aug–Oct) with many overwintering ✦ Abundant (hundreds of thousands) especially in Mediterranean regions; rare in the humid N and mountains. Common in winter in Extremadura, Andalucía and the Mediterranean coastlands. Common in Balearics and Canary Is.

CORACIIFORMES (families Alcedinidae, Meropidae, Coraciidae) – Brightly-coloured, arboreal or terrestrial nesters, with partly fused toes. Generally insectivorous. Nidicolous. Kingfishers worldwide, others Old World. Mainly tropical.

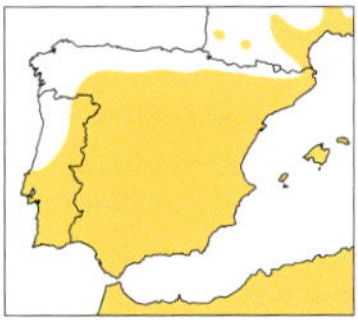

European Bee-eater *Merops apiaster* (abejaruco europeo). ~28 cm. Distinctive call: a far-reaching, whistled, liquid *prriip* or *trruic*. Slim, with long, pointed wings and tail. Coloration exotic. Female greener above on wings and rump. Juvenile duller; lacks elongated central rectrices. Hunts insects in flight, often in flocks. Often perches on wires and fences ✦ Widespread in open country; absent from forests and high elevations. Nests colonially, digging burrows in sandbanks ✦ Winters in Africa (passage ~Apr and Aug–Oct) ✦ Very common (many hundreds of thousands) throughout the Mediterranean region, abundant in many parts of C and S. Scarce migrant in Canary Is.

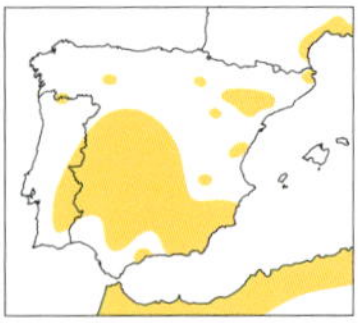

European Roller *Coracias garrulus* (carraca europea). ~31 cm. Beautifully colourful when seen at close range; often perches on posts and cables. Size of Stock Dove or Jackdaw; could be mistaken for them at long range. Bill heavy, corvine. Juvenile duller, yellowish-toned on head. Has tumbling display flight, when calls like a Magpie. Flies down to prey from perches ✦ Open landscapes or those with some trees, in dry regions with pastures, fallows, dry crops. Nests in holes in trees, cuttings or small buildings; often in nest boxes ✦ In summer and on passage (Apr/May–Aug/Sept) ✦ Uncommon (a few thousand, declining); frequent in Extremadura and arid SE (*ramblas*). Rare migrant in Balearic and Canary Is. **VU**

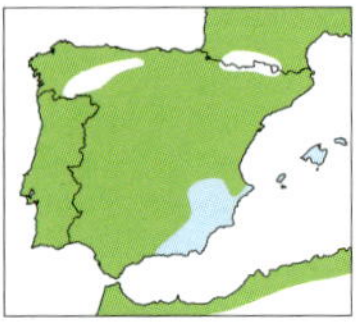

Common Kingfisher *Alcedo atthis* (martín pescador común). ~17 cm. Frequently appears as a turquoise-blue flash over a river, accompanied by a penetrating whistled *chreee*. Often perches on twigs over water. Head and bill large; legs very short; brightly coloured. Juvenile duller. Dives for fish from a perch, or after hovering ✦ Clear streams and rivers with marginal vegetation and sandbanks for nest burrows; avoids mountain streams and polluted waters. In winter visits irrigation channels, pools, reservoirs, lakes, estuaries ✦ Spanish birds sedentary, with juvenile dispersion. Wintering (~Sept–Feb) and passage of birds from France and nearby countries ✦ Ssp. *ispida* (with transition to *atthis* in E and S): quite common (a few thousand pairs, declining); densities vary according to water quality and levels; rare in arid SE. Scarce in Balearics. Vagrant to Canary Is. **NT**

Common
Hoopoe
European
Bee-eater
European Roller
Common Kingfisher

WOODPECKERS (Piciformes, Picidae) – Creepers with two toes forward and two back. Have extremely long tongues to extract insects and their larvae. Often on tree trunks and branches. Size very variable. Nest in tree holes that they often excavate; young nidicolous. Generally in woodlands. Eurasia, Africa and the Americas, in a wide range of climates.

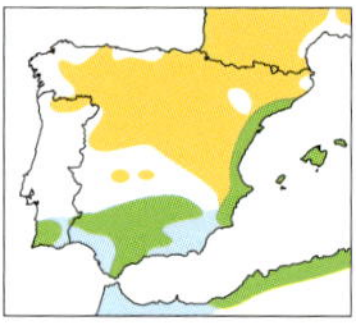

Eurasian Wryneck *Jynx torquilla* (torcecuello euroasiático). ~17 cm. Highly distinct from other woodpeckers. Has size, appearance and behaviour of a passerine. Highly cryptic plumage recalls nightjars. Often detected by call: a repeated *ki, ki, ki, ki, ki…* suggesting a small falcon but more monotonous and less urgent, like Lesser Spotted Woodpecker; migrants also call. Bill short, pointed; dark stripes on centre of back, scapulars; dark eye stripe descends neck; crown feathers erectile. Discreet; hunts ants on ground; perches on twigs but can also creep like a woodpecker ✦ Wooded countryside, orchards (orange groves, apple orchards, carobs), riparian woods, ash groves, open woodlands; generally below 1.300 m. Requires tree holes for nesting ✦ Transaharan migrant, with passage of European birds (mainly Mar–Apr and Sept–Nov) and many over-winterers ✦ Quite common breeder (some tens of thousands of pairs, trends varying between regions); well distributed in NE quadrant but ever scarcer towards the W and S, where typically restricted to cooler montane areas but now also spreading there. Scarce in winter, then mainly in Andalucía and the Mediterranean coastlands. Breeds and winters in the Balearics. Scarce and irregular on passage in Canary Is. **DD**

Iberian Green Woodpecker *Picus sharpei* (pito real ibérico). ~32 cm. The only green woodpecker in Spain, until recently treated as a race of the **Eurasian Green Woodpecker** *Picus viridis* (pito real europeo) which breeds from France onwards. Call: a ringing *kik–kik–kik–kik–kik–kik*, recalling a neighing horse; rarely drums. Greenish; yellow rump obvious in flight; striking red cap; moustachial stripe red in male and black in female. Juvenile duller and very spotted. Flight undulating, folding wings between bursts of wingbeats. Often on the ground, seeking ant nests ✦ Varied habitats, from deciduous or coniferous woods with clearings to very open countryside with few trees; often in riparian woods, orchards and parks. Excavates nest hole in a tree; sometimes in an earth bank ✦ Sedentary ✦ Very common (up to a few hundred thousand pairs) and widespread, although most frequent in N and C. Scarce in treeless regions and absent from large parts of Extremadura and Andalucía, as well as from Ceuta and Melilla.

Black Woodpecker *Dryocopus martius* (picamaderos negro). ~45 cm. Pyrenees and Cantabrian range. Crow-size and all black except for crown; all-red in male, red on hind-crown only in female. Bill and iris whitish in adults. Juvenile browner, with crown duller; iris blue-grey. Flight call: a loud *krrri–krrri–krrri*. Song recalls Iberian Green Woodpecker but slower: *kui, kui, kui…*; drums very loudly, in bouts of ~30 blows per 2 s. Flight level with irregular wingbeats, undulating only when landing ✦ Mature montane forests, chiefly at 1,000–1,700 m (beech, spruce, black and Scot's pine, oaks) ✦ Sedentary; some juveniles disperse far ✦ Scarce (hundreds of pairs, now increasing) and local in northern forest blocks; highest densities in Pyrenees of Catalonia and Aragón. There are very rare records from Central range and elsewhere. Latterly seems to be expanding range, as elsewhere in W Europe.

Eurasian Wryneck
Iberian Green Woodpecker
♀
Black Woodpecker
♀

Great Spotted Woodpecker *Dendrocopos major* (pico picapinos). ~23 cm. Common. Black above with clear, distinct white areas on face, collar and scapulars; dirty-white below, vent area bright red. Black moustachial stripe joins other black lines to nape and breast-sides. Cap black, with red nape patch in male; all-red in juvenile. Flanks unstreaked in adult, may be finely streaked in juvenile. White scapulars and barred tail-sides evident in flight. Call: a sharp *tchik*. Often drums, in short bursts of 12 blows per s ✦ All woodland types; also in riparian woods and parks, sometimes poplar plantations; favours montane pinewoods and oakwoods ✦ Sedentary ✦ Ssp. *hispanus*: common (several hundred thousand), abundant in wooded regions but scarce in mesetas, Ebro valley and arid SE. Vagrant to Balearic Is. Ssp. *canariensis*: Tenerife (hundreds, perhaps increasing), **VU**. Ssp. *thanneri*: Gran Canaria (hundreds of pairs), **VU**

White-backed Woodpecker *Dendrocopos leucotos* (pico dorsiblanco). ~26 cm. Navarran Pyrenees. Some 10% larger than Great Spotted Woodpecker and lacks white scapulars; instead regularly barred white on back and wings. Facial pattern similar to Great Spotted but cap all red in male, black in female and mixed in juvenile. Flanks very streaked. Vent area strong pink, intermediate in intensity between vent colour of Great Spotted and Middle Spotted Woodpeckers. Call more muted than Great Spotted: *kiik* but drums much louder and longer (~30 blows in 2 s) ✦ Mature montane forests, chiefly beech or spruce (Irati, Quinto Real) ✦ Sedentary ✦ Ssp. *lilfordi*: rare (tens of pairs) and very local; perhaps recently expanding range west. **VU**

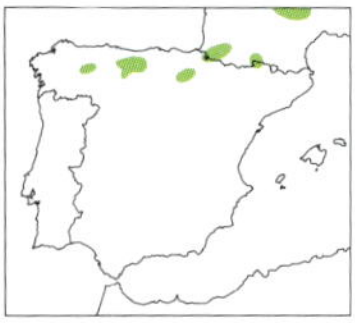

Middle Spotted Woodpecker *Leiopicus medius* (pico mediano). ~21 cm. In N only. Scapulars white, as in Great Spotted Woodpecker, but some 10% smaller and head whiter; moustachial stripe short and incomplete, neck stripe does not reach nape. Cap always red; vent pink (not red). Yellowish below, flanks streaked. Seldom drums but often calls in short series: *kíg–gag–gag–gag–gag*. Territorial call a nasal, jay-like *gwaaaag* ✦ Mature oakwoods ✦ Sedentary ✦ Scarce (hundreds of pairs) and local, concentrated in Cantabrian range (provinces of León, Palencia, Asturias and Cantabria) and in Pyrenees in Álava/Navarra. **NT**

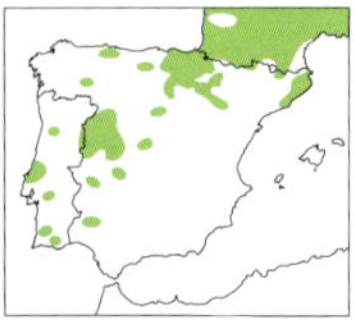

Lesser Spotted Woodpecker *Dryobates minor* (pico menor). ~15 cm. Nuthatch-sized. Barred above, lacking white scapulars. Whitish below, vent included. Flanks streaked. Bill short, complete black moustachial stripe. Cap red in male, black in female/juvenile. Call: *ki–ki–ki–ki*…, as Wryneck but softer, or like a distant Kestrel; in series of 8–20 notes, at 6 notes per s. Drum weaker but longer than Great Spotted Woodpecker; 20–30 blows in 1.5 s. Often among high, slender branches, where easily overlooked ✦ Oakwoods, mature dehesas, riparian woods, alder carr, chestnut and poplar plantations, etc., selecting mature trees ✦ Sedentary, with juvenile dispersal ✦ Ssp. *buturlini*: quite scarce (a few thousand pairs) and local, with principal concentrations in Basque Country and W of Central range (Cáceres and Salamanca provinces); some recent range expansion, at least into Catalonia, the Ebro valley and Madrid province.

♂ ad.
Great Spotted Woodpecker
♂ juv.
♀ ad.
♂ ad.
White-backed Woodpecker
Middle Spotted Woodpecker
♂ ad.
Lesser Spotted Woodpecker

FALCONS (Falconiformes, Falconidae) – Small to medium-sized, compact raptors. Wings long and pointed. Bills short, 'toothed' on upper mandible. Close relatives of parrots and passerines. Worldwide.

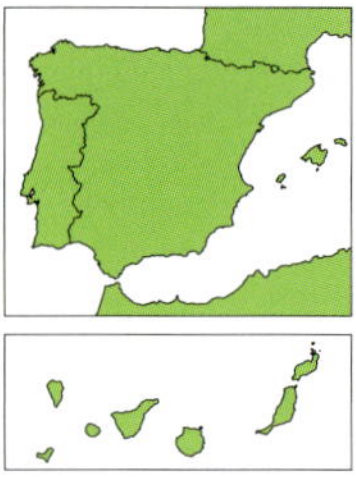

Common Kestrel *Falco tinnunculus* (cernícalo vulgar). 32–35 cm. Ubiquitous, long-tailed, often hovers. Quite similar to Lesser Kestrel (below). Sparrowhawk (p. 134) has much broader, rounded wings. Male has head grey, moustache black; back chestnut, spotted black. Female brown, densely barred and spotted. Juvenile resembles female; more streaked below. Call: a sharp *kee–kee–kee* ✦ Broad habitat selection; nests on cliffs, trees (using corvid nests), buildings, etc. Hunts in croplands, pastures, scrub, roadside verges ✦ Resident, with seasonal and dispersive movements. Some numbers of European birds occur on passage and in winter (Aug/Nov–Feb/May) ✦ Very common (some tens of thousands) and widespread. Many winterers in Mediterranean coastlands. Abundant in Canary Is (some thousands of pairs): ssps. *dacotiae* (Lanzarote and Fuerteventura), **VU**, and *canariensis* (elsewhere), **DD**

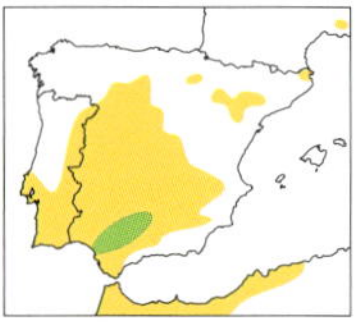

Lesser Kestrel *Falco naumanni* (cernícalo primilla). ~30 cm. Towns, villages and outskirts. Smaller, paler, more delicate and more gregarious than Common Kestrel and wingbeats faster. Claws white. Underwings much whiter; central rectrices project (esp. in males); wingtips reach tail tip at rest. Adult male has head bluish, lacks moustache; brick red above, unspotted; broad grey-blue band on greater coverts. Female and juvenile less strongly marked than in Common Kestrel, with weak moustachial stripe. Call characteristic: a sibilant, trisyllabic, *chet–chet–chet* ✦ Colonial; colony size very variable; nesting in holes in walls or under roof tiles on churches, convents, castles, ruins, byres, old buildings. Very few now nest on cliffs. Hunts insects over pastures, fallows, cereal crops ✦ Summer visitor (Feb/Apr–Sept/Oct); some numbers of adults overwinter in S ✦ Some 20,000 pp., now increasing. Common or very common in W Andalucía (where winters), Extremadura and parts of Castilla-La Mancha and Aragón (Monegros), rarer towards N and E. Occasional in Balearics. Vagrant to Canary Is. **VU**

Red-footed Falcon *Falco vespertinus* (cernícalo patirrojo). ~30 cm. Rare migrant. Structure between Hobby and Kestrel. Adult males distinctive. More frequent 1st summer males show whitish face, orange breast patches and barred tail. Females have dark face mask; dark upperparts contrast with orange-buff underparts; 1st summer females browner with streaked cap, nape and flanks. Legs red or reddish. Solitary or in small groups. Often perches on posts and cables; captures insects in flight; hovers ✦ Open areas, often near wetlands ✦ On passage (Apr–Jun and Sept–Oct); nearest breeding areas in Hungary and Romania ✦ Scarce, numbers very variable between years; annual in spring in coastal Catalonia and the Balearics; much rarer in autumn. Vagrant to Canary Is.

Merlin *Falco columbarius* (esmerejón). 25–30 cm. Size of turtle dove; compact and unshowy. Adult male (relatively infrequent in Spain) bluish-grey above, orangey below, with black terminal tailbar. Female and immature dark brown above, heavily streaked below, tail densely barred. Solitary. Often perches on the ground or a rock, where easily overlooked. Rapid and agile in flight, often skimming ground, pursues small birds determinedly ✦ Agricultural land, pastures, dwarf shrub steppes, wetlands; avoids mountains and forests ✦ In winter (Sept/Nov–Feb/Apr), nesting in Scandinavia, Finland and British Isles (ssp. *aesalon*) ✦ Uncommon (many hundreds to a few thousand), most frequent in N of Castilla and in Ebro valley. Rare in Balearics. Vagrant to Canary Is.

Common Kestrel
♂
♀
Lesser Kestrel
♂
♀
Red-footed Falcon
♂
♀
Merlin
♂
♀

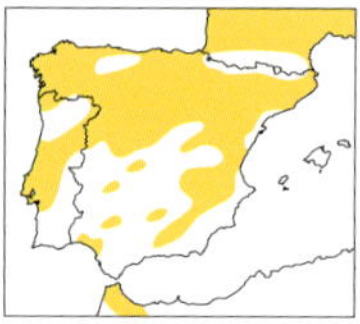

Eurasian Hobby *Falco subbuteo* (alcotán europeo). 30–35 cm. Slim, aerial falcon; long-winged, wingtips project beyond tail at rest. Dark. Streaked below. Adult has chin and cheeks very white; 'trousers' and vent red. Juvenile lacks red areas; browner than adult with pale feather edging above. Hunting flight relatively slow when hunting insects, often at dusk, but pursues birds, often hirundines, at high speed. Call: *kew–kew–kew–kew* ✦ Open terrain with some trees, copses or woodland edges (pinewoods, oakwoods, holm oak woods, riparian trees). Uses corvid nests ✦ In summer and on passage (Apr/May–Sept/Oct) ✦ Uncommon (a few thousand pairs), most frequent in N half, especially in Duero basin; scarce and local in S half. Uncommon in Balearics on passage (has bred Mallorca). Scarce and irregular in Canary Is. **NT**

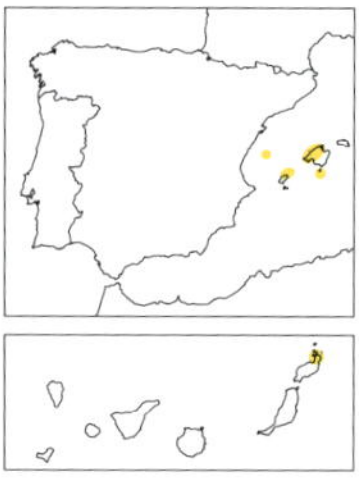

Eleonora's Falcon *Falco eleonorae* (halcón de Eleonora). 36–40 cm. Nests on islands. Approaches Peregrine in size but has slim build of Hobby, although longer-tailed. Two morphs: dark (~25%) and light, latter resembling a Hobby with rufous-buff underparts. Both show contrast between dark underwing coverts and paler remiges. Juvenile browner; remiges and rectrices strongly barred. Flies in flocks around sea cliffs ✦ Nests in small colonies, on sea cliffs on islands and islets, feeding on migrating landbirds. Earlier hunts insects over pinewoods, scrub, wetlands, crops and even towns ✦ Summer visitor (Apr–Oct/Nov), nesting very late (Jul–Oct). Winters in Madagascar and E Africa ✦ Uncommon (some hundreds of pairs, increasing). Breeds locally in Balearics (some 500 pp., in Serra de Tramuntana, Dragonera, Cabrera, Ibiza), Columbretes (tens) and Canary Is (some 200 pp. on islets N of Lanzarote). Regular, although scarce, in littoral areas of Catalonia/Valencia (mainly May–Aug); also at times further inland (around Jun–Jul). **NT**

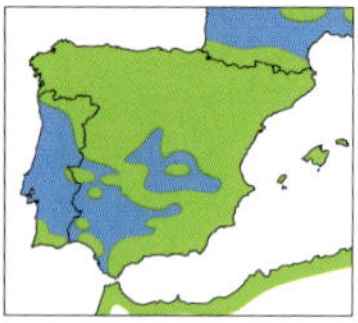

Peregrine Falcon *Falco peregrinus* (halcón peregrino). 40–45 cm. Powerful, compact, deep-chested. Wings broad but pointed, tail relatively short. Adult whitish or pinkish-tinted below, blue-grey above; head black and white with broad black moustaches. Female substantially larger. Juvenile brown above, heavily streaked, not finely barred, dark below; moustaches less clear-cut. Flight very rapid; overtaking or stooping on prey: birds caught in flight. Calls noisily in alarm near nest: *keyak–keyak–keyak* ✦ Nests at very variable altitudes; on sea cliffs, gorges, mountain cliffs, towers, castles and other tall buildings, in corvid nests on electricity pylons, etc., generally alongside open terrain or cultivation. Increasingly often in cities ✦ Spanish birds sedentary, with juvenile dispersion. Wintering and passage of some European birds (Oct–Apr) ✦ Ssp. *brookei*, breeds quite commonly (two or three thousand pairs, increasing gradually) and widespread, although dependent on nest-site availability (hence scarcer in plains of Castilla-La Mancha, Extremadura and Andalucía). Ssps. *peregrinus* and perhaps also *calidus*: in winter and on passage; scarce. Scarce and irregular in winter in Canary Is. Ssp. *pelegrinoides*, **Barbary Falcon** (halcón tagarote), sometimes treated as distinct species; smaller, less bulky, paler above and poorly marked below; rear crown and nape rufous: solely in Canary Is, on sea cliffs and some escarpments inland, where scarce (tens of pairs, increasing) although quite widespread. **EN**

Eurasian Hobby
juv.
Eleonora's Falcon
♂ pale morph
♂ dark morph
juv.
Peregrine Falcon

PARAKEETS (Psittaciformes, Psittacidae) – Arboreal fruit- and seed-eaters, with short hooked beaks and typically bright green plumage. Feet have two toes forward, two back. Exotic; of captive origin in Europe; two species are naturalised in Spain and other escaped species may occur.

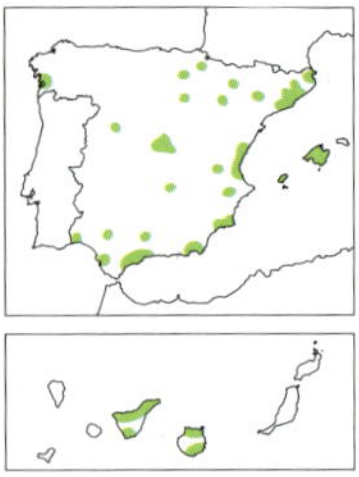

Monk Parakeet *Myiopsitta monachus* (cotorra argentina). ~30 cm. South American. Similar in size to Rose-ringed Parakeet but shorter tailed. Predominantly green, with grey throat and breast; bluish remiges evident in flight. Bill pale pinkish-brown. Forehead grey in adult, greenish in juvenile. Contructs large communal stick nests shared by several pairs; on palms, cedars, eucalypts, electricity pylons, etc. Flocks very noisy ✦ Parks, tree-lined avenues, golf courses ✦ Locally common (a few thousand pairs, rapidly increasing; first appeared in 1970s and 1980s). Most evident in Barcelona and Madrid but also in Valencia, Málaga, Cádiz, Palma de Mallorca, Canary Is, etc.

Rose-ringed Parakeet *Psittacula krameri* (cotorra de Kramer). ~40 cm. From Africa and S Asia. Very long-tailed. Bright green; beak has upper mandible dark red, lower black. Adult male has narrow pink and black collar, absent or pale green in female, lacking in juvenile. Loud shrieking call striking, although less strident than Monk Parakeet. Nests in tree holes ✦ Parks, golf courses, etc. ✦ Sedentary ✦ Scarce or uncommon (some hundreds of pairs) but increasing; chiefly in Madrid, Barcelona, Valencia, Alicante, Málaga, Mallorca, Canary Is.

ORIOLES (Passeriformes, Oriolidae) – Medium-sized arboreal birds, frugivorous and insectivorous. Allied to corvids and shrikes. Chiefly in Old World tropics and Australia.

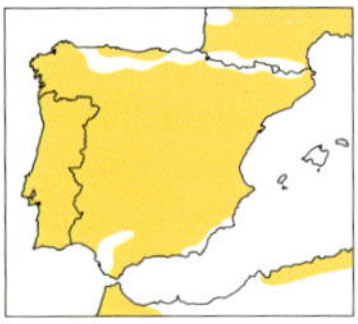

Eurasian Golden Oriole *Oriolus oriolus* (oropéndola europea). ~24 cm. Often revealed by its song, a fluting *weela–weeoo*. Call: a harsh, mewing *weeach.* Keeps in tree canopy; shy and hard to see. Male brilliant yellow. Female and juvenile greenish, streaked below; confusable with Green Woodpecker but much smaller, thrush-like ✦ Riparian woodlands, poplar plantations, woodlands (especially oak); at low elevations. Slings nest across a forked branch ✦ Trans-Saharan migrant (late Apr/May–Aug/Sept) ✦ Quite common (some hundreds of thousands of pairs) and well distributed, although scarce in northern coastlands and avoids higher elevations. Scarce passage migrant in Balearics. Scarce and irregular in Canary Is (around Apr–May).

BUSH SHRIKES (Passeriformes, Malaconotidae) – Resemble true shrikes, to which allied. Head large with strong, hook-tipped beak. Hunt insects and small vertebrates. Africa.

Black-crowned Tchagra *Tchagra senegalus* (chagra del Senegal). ~22 cm. N Africa. Head pattern unmistakable; black crown and eye stripes contrast with broad white supercilium. Wings rufous; tertials and scapulars blotched black. Tail long, dark, white-tipped. Bill large, strong. Sexes similar. Generally skulks in dense cover but may hop around in open like a magpie. Frequently revealed by loud melodious song, a series of wavering whistles, often duetting with female ✦ Scrub and open woodland (kermes oak, lentisc, wild olive, juniper), also in tamarisks and oleanders along streams and riverbanks ✦ Resident, with postbreeding dispersal ✦ Ssp. *cucullata*: local in coastal regions of Morocco and Algeria; seen occasionally in Ceuta (one Peninsular record from Tarifa, S Andalucía).

Monk Parakeet
Rose-ringed Parakeet
Eurasian Golden Oriole
♂
♀
Black-crowned Tchagra

SHRIKES (Passeriformes, Laniidae) – Large-headed, with strong, hooked beaks. Hunt insects and small vertebrates from perches. Eurasia, Africa, New Guinea and N America.

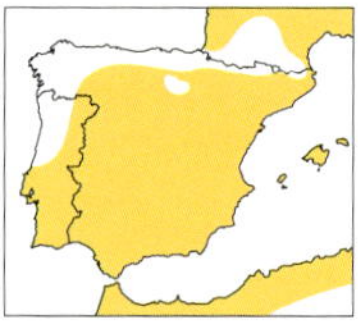

Woodchat Shrike *Lanius senator* (alcaudón común). ~18 cm. Summer visitor. Cap and nape rufous. Black and white, with broad white patches on scapulars, rump and primary bases. Black mask extends to forehead. Female duller, often buffy-white around eye. Juvenile lacks mask, grey-brown, dark feather tips give vermiculations below and a scaly effect above (resembles juvenile Red-backed Shrike but less rufous and shows pale scapular patches and pale rump contrasting with dark tail). Vocalisation diverse. Call: a harsh *ki-wick ki-wick*. Song loud, rich and varied, with warbling, whistles and mimicry. Perches conspicuously, often on wires and posts, solitary ✦ Open country with some trees: scrub clearings and edges, dehesas, riparian woodlands, olive and almond groves, etc. ✦ Trans-Saharan migrant (late Mar/Apr–Aug/Sept) ✦ Ssp. *rutilans*: common or very common (hundreds of thousands of pairs), but rapidly declining; most abundant in S and SW. Ssp. *badius*: very common in Balearics. Ssp. *senator*: perhaps a scarce migrant (from France). Ssp. *niloticus*: vagrant (Asia). The species is regular on passage in Canary Is. **NT**

Red-backed Shrike *Lanius collurio* (alcaudón dorsirrojo). ~17 cm. Male has black mask; rufous above, with grey crown, nape and rump. Female, usually much browner and duller; vermiculated below. Wings unmarked; tail shows white bases on sides. Juvenile brownish, barred above; rump and crown rufous, ear coverts dark, wings lack white. Sings rather seldom; quietly; a prolonged swallow-like twittering; also mimetic. Call: *check* ✦ Northern countryside with hedges and meadows; spiny scrub at woodland edges; montane pastures with shrubs ✦ Summer visitor (late Apr/May–Aug/Sept) migrating SE (imperceptibly) ✦ Common (a few hundred thousand pairs), very common in Cantabrian mountains and Pyrenees, scarcer in Galician coastlands; spreading southwards. Occasional outside breeding range. Rare in Balearics. Vagrant to Canary Is.

Lesser Grey Shrike *Lanius minor* (alcaudón chico). ~20 cm. NE; virtually extinct. Resembles a small Southern Grey Shrike with mask extending as a broad band across forehead; tail shorter and less rounded. Beak stout. Strongly pink tinted below. Juvenile lacks black forehead; shows 'scaly' pattern above ✦ Open countryside with fallows and trees ✦ Trans-Saharan migrant (May–Aug) migrating SE ✦ Very rare and local (SW Lleida), reintroduction of captive bred birds in progress; **CR**. Vagrant to Balearic Is.

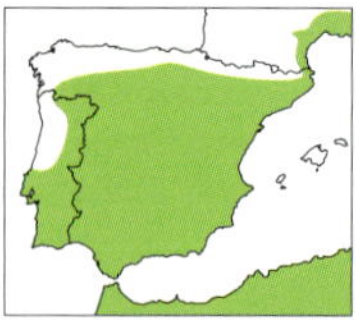

Southern Grey Shrike *Lanius meridionalis* (alcaudón real). ~24 cm. Year-round. Large, dark grey above, pinkish-tinted below. Treated as a race of Great Grey Shrike until recently. Narrow white supercilia meet on forehead. Tail long, rounded. Juvenile paler above, browner, finely barred below. Flight strong, undulating. Often perches on wires ✦ Open country with trees or scattered shrubs: fallows, pastures, dehesas ✦ Resident, with dispersal ✦ Ssp. *meridionalis*: quite common (a few hundred thousand), most frequent in C and SW but declining strongly; rare winterer in Balearics; **NT**. Ssp. *koenigi*: Canary Is, uncommon (many hundreds of pairs), most frequent on eastern islands; **DD**

Great Grey Shrike *Lanius excubitor* (alcaudón norteño). ~25 cm. In winter. Paler than Southern Grey Shrike and lacks rosy tint below; shows white on secondary bases and broader white edging to scapulars; white supercilia less evident ✦ In winter (around Oct–Mar). Nests from France to Scandinavia and across N Asia and N America ✦ Selects open country ✦ Status uncertain but probably a scarce winter visitor to the N. Vagrant to Balearic Is.

♂ ad.
juv.
Woodchat Shrike
juv.
♀ ad.
Lesser Grey Shrike
♂ ad.
Red-backed Shrike
Southern Grey Shrike
Great Grey Shrike

CORVIDS (Passeriformes, Corvidae) – Relatively large or very large passerines. Have robust beaks with rictal bristles and large feet with strong claws. Plumages vary; sometimes all black and often iridescent. Sexes similar. Allied to shrikes and orioles. Adaptable omnivores, occupying a great diversity of habitats. Worldwide.

Eurasian Jay *Garrulus glandarius* (arrendajo euroasiático). ~34 cm. Woodlands. Pale pinkish brown; in flight shows white secondary patches and striking white rump. Wings rounded; tail longish, black. At close range shows strong black moustachial stripe, streaked cap and blue, black-barred patch on wing bend. Call: a harsh *schaach*, often repeated. Flight weak, undulating. Quite shy and wary ✦ Woodlands, especially deciduous oak woods (also holm oak, beech and pine woods), especially in montane areas. Seeks and buries acorns in autumn ✦ Sedentary ✦ Ssps. *glandarius* (N), *lusitanicus* (NW) and *fasciatus* (S and E): common (many hundreds of thousands of pairs), probably increasing; most frequent in N and mountains; absent from treeless areas. Vagrant to Balearic Is.

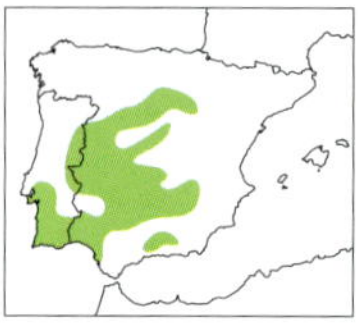

Iberian Azure-winged Magpie *Cyanopica cooki* (rabilargo ibérico). ~34 cm. Formerly regarded as a subspecies of the **Asian Azure-winged Magpie** *C. cyanus* (rabilargo asiático) native to China, Korea and Japan. Shape like a small Common Magpie but coloration very different and beats wings steadily in flight; black cap contrasts with white throat; body brownish-pink; tail and wings pale blue. Juvenile duller. Often in noisy flocks; calling frequently: *zhreee*. Hops on ground. Restless and inquisitive. Often nests in loose colonies ✦ Open Mediterranean woodlands (holm oaks, pines, junipers), dehesas, orchards, etc. ✦ Sedentary, with juvenile dispersal ✦ Common (hundreds of thousands of pairs) apparently increasing; abundant in wooded areas of Extremadura (especially Cáceres) and neighbouring regions (Sistema Central, Sierra Morena); scarcer in Duero basin and Granada/Málaga; occasionally observed elsewhere.

Common Magpie *Pica pica* (urraca común). ~45 cm. Very common and familiar. Unmistakable black and white plumage with long graduated tail, showing green and purple iridescence in good light. Flies with alternating series of wing beats and glides, wings rounded, relatively short, showing much white on primaries. Call also familiar: a rapid, harsh, rattling *chac–chac–chac–chac*; has other, unmusical, calls. Corvine beak; legs quite long. May show some bare blue skin behind eye at times. Walks and hops readily on the ground, often with tail raised. Solitary, in pairs or in small flocks. Builds large domed stick nests in trees or large shrubs ✦ Uses a wide variety of habitats, selecting croplands and outskirts of towns (riverine trees, orchards, parks); avoids mountains, forests and treeless regions. Ever commoner in urban habitats. Eats insects, fruits, seeds, small vertebrates, carrion and garbage ✦ Sedentary ✦ Ssp. *melanotos*: very common (many hundreds of thousands or up to a few million birds); widespread and spreading, although still absent from parts of Andalucía and the SE. Possible vagrant to Mallorca. Much persecuted by game hunting interests.

Eurasian Jay

Iberian Azure-winged Magpie

Common Magpie

Rook *Corvus frugilegus* (graja). ~45 cm. Very local. Differs from Carrion Crow in having an area of bare, greyish-white skin on the face and around bill base; absent in juvenile and 1st year. Steep profile, shaggy flanks and dense neck plumage characteristic; bill relatively slender and pointed. Wings somewhat narrow and pointed. Tail very rounded, almost wedge-shaped. Call: *kaah*, less raucous and more nasal than that of Carrion Crow. Highly gregarious year-round ✦ Nests in rookeries; colonies of large nests in isolated poplar plantations, often alongside villages and near fields and meadows; latterly also in town parks. Feeds on farmland. Visits rubbish dumps ✦ Breeding birds sedentary, with short-distance dispersion. A few European birds over-winter (Oct–Mar) ✦ Breeding birds scarce (over 1,000 pairs, fluctuating, and restricted to León province (especially between León city and La Bañeza). Winter immigrants once very common and well distributed, especially on the Meseta and in the Ebro valley, but now largely remain further north in Europe. Winter records away from the breeding area now only sporadic. Vagrant to Balearic Is. **VU**

Carrion Crow *Corvus corone* (corneja negra). ~47 cm. Some 25% smaller than Raven; bill less massive, tail straight-ended. Larger than Jackdaw and lacks grey nape. (See above for distinction from Rook). Grating rolling call: *crraa*, less deep than Raven; has various other vocalisations. Often alongside roads, frequently perched on posts. Forms family parties. Replaced over much of Europe by the **Hooded Crow** *C. cornix* (corneja gris), a vagrant to Spain, chiefly to the NE and NW (p. 235) ✦ Countryside with trees and copses near cultivation. Nests in trees ✦ Sedentary but dispersive ✦ Common (hundreds of thousands of pairs) but irregularly distributed and densities very variable: very common in N half (Duero basin, Biscay coastlands and Galicia) but absent from much of S half (most of Extremadura, Andalucía, Murcia and Mediterranean coastlands). Much persecuted by game hunting interests. Vagrant to Balearic Is.

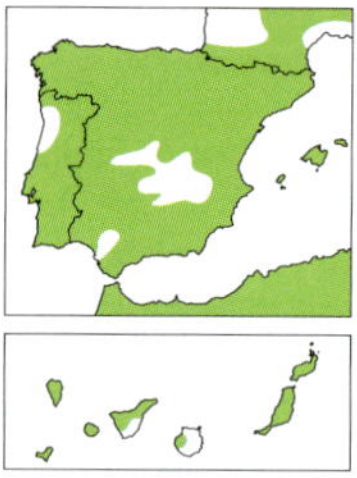

Common Raven *Corvus corax* (cuervo grande). ~60 cm. Largest passerine, buzzard-sized. Bill massive; tail long, wedge-shaped; throat hackles shaggy. Vocalisations deep and hoarse; including a brief *prark*. Flight powerful with regular wingbeats and long glides and rolling displays; soars, showing characteristic 'Maltese cross' silhouette; chases off raptors. Often solitary or in pairs and family parties, but sometimes also in large flocks (especially at rubbish dumps, roosts, carcasses) ✦ Forested or open terrain at any altitude, most frequent in mountains and remote areas. Nests on cliffs, ruins, trees, electricity pylons ✦ Sedentary, but dispersive ✦ Ssp. *hispanus*: relatively common (many tens of thousands of pairs) and well distributed, scarce only in treeless plains, especially in the S half. Common in Balearics. Ssp. *tingitanus*: nests in Ceuta. Ssp. *canariensis*: Canary Is, a few hundred pairs, declining, very few now in Tenerife and Gran Canaria; **EN**

Rook
Carrion Crow
Common Raven

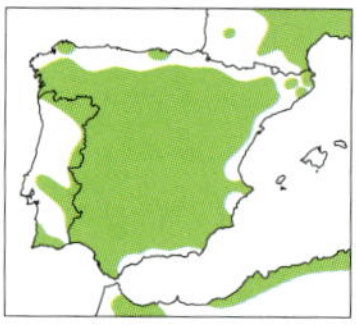

Eurasian Jackdaw *Corvus monedula* (grajilla occidental). ~32 cm. Smallest black corvid; bill fairly short, nape grey, face dark, iris pale (dark in juvenile). Gregarious, noisy; typical call: *kyack kyack*, recalling Red-billed Chough but shorter and harsher. Compact in flight, wings relatively narrow and pointed; wingbeats faster than crows. Attends rubbish dumps. Flocks tend to remain near breeding sites all year ✦ Habitat varies but influenced by nest site availability. Nests in cavities on cliffs, historic buildings (often in towns and cities), ruins, stone bridges, trees with holes. Favours farming areas, especially cereal croplands, and avoids forests and high mountains ✦ Spanish birds resident and dispersive. Small numbers of European birds over-winter (Oct–Mar) ✦ Ssp. *spermologus*: common (hundreds of thousands of pairs), declining and still quite widespread, with high densities in some areas, but scarce or local in Galicia and the Biscay coastlands; declining rapidly in Catalonia. Occasional in the Balearics.

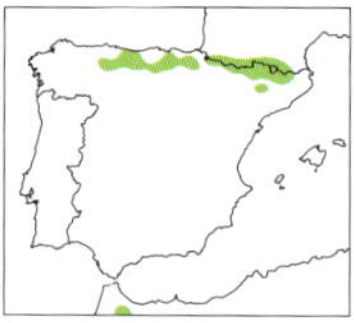

Yellow-billed Chough *Pyrrhocorax graculus* (chova piquigualda). ~38 cm. Northern mountains. Bill yellow (browner in juvenile), shorter and less decurved than in Red-billed Chough. Calls include a rippling *preeep* and a descending, whistling *sweeeeooo*. Less glossy than Red-billed and forehead feathering extends further on to upper mandible; tail rounded, longer than wing width and extends beyond wingtips at rest. Comparatively short-legged. Wing tips more rounded and less 'ragged' than in Red-billed. Highly gregarious, often in large flocks. Superb flier ✦ Nests in alpine or alpine-like zones, generally above 1,000 m. Nests in cliffs, caves and potholes. Hunts invertebrates on pastures and scree; often attends ski stations ✦ Sedentary, with altitudinal movements in winter, sometimes to nearby ranges ✦ Quite common (a few thousand pairs) but only nests in the Pyrenees and Cantabrian range; a little more widespread in winter; locally very common (in karstic zones); occasional records from Andalucía suggest dispersion from Morocco. Vagrant to Balearic Is.

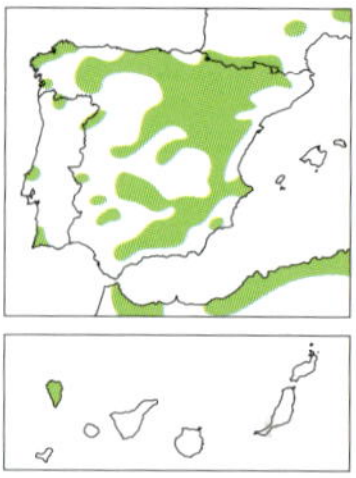

Red-billed Chough *Pyrrhocorax pyrrhocorax* (chova piquirroja). ~40 cm. Widespread. Like a Jackdaw, with red legs and long, red, decurved bill; all black, glossy. Wings broad, squarish, with deeply separated primaries; tail shorter than wing width. Call: the eponymous *chuff*, louder and longer than Jackdaw call. Juvenile has shorter, more orange bill: sometimes confused with Yellow-billed Chough (see above). Highly acrobatic in flight. Gregarious, with mate fidelity ✦ Nests colonially or as isolated pairs on cliffs and escarpments, both coastal and inland, and often on ruins, historic buildings, etc. Often ascends to mountain pastures, where both chough species may forage together. Searches for insects on level pastures, often among open scrub, or in croplands and fallows ✦ Sedentary, with altitudinal movements ✦ Ssp. *erythrorhamphos*: population fairly small (up to a few tens of thousands) but widespread; most frequent in limestone areas of the E half. Vagrant to Balearic Is; **NT**. Ssp. *barbarus*: only on La Palma, Canary Is, where common (over 1,000 birds), **EN**

Eurasian Jackdaw
Yellow-billed Chough
Red-billed Chough

TITS (Passeriformes, Paridae) – Small, acrobatic, insectivorous woodland birds. Gregarious, often in mixed species flocks. Bill short, legs strong. Also eat nuts, fruits and seeds. Nest in holes and nestboxes. Sexes similar. Eurasia, North America and Africa.

Great Tit *Parus major* (carbonero común). ~14 cm. Considerably larger than the Blue Tit, the only other tit with yellow underparts; head black with white cheeks, lacking blue or a supercilium; black median breast and belly stripe. Male brighter yellow than female with broader median stripe, on belly particularly. Juvenile duller, face yellowish. Calls variable, some resembling those of Blue Tit but less sharp. Include Chaffinch-like *pink*. Song repertoire includes bi- and trisyllabic phrases, repeated: *tee–cher, tee–cher* and *chee–chee–cher, chee–chee–cher* among others. Inquisitive, confiding ✦ Any areas with trees (forests, dehesas, riparian woods, orchards, gardens, etc.), at any altitude, although less frequent in conifers ✦ Resident, with short-range dispersive and altitudinal movements; some immigrants from other European countries ✦ Ssps. *major* (N) and *corsus* (S): very common (millions of pairs) and increasing; widespread. Ssp. *mallorcae*: Balearics, common, often in pinewoods. Ssp. *excelsus*: breeds Ceuta and Melilla. possible vagrant to Canary Is.

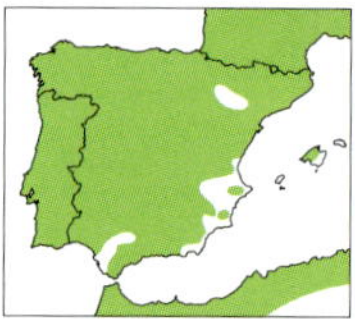

Common Blue Tit *Cyanistes caeruleus* (herrerillo común). ~11 cm. Small, very attractive; blue on cap, wings and tail. Face white with clear dark eyestripe. Female less bright; juvenile duller, lacks bib, face also yellow. Vocalisation: song *chi–chi cheeeeer* (last note harsher), alternating *tsi–tsí* and others. Confiding ✦ All types of woodlands and dehesas, although tends to avoid conifers, selecting broadleaves with old trees; also riparian woods, parks, gardens, open country with trees ✦ Sedentary, with short-range movements and irruptions of European birds ✦ Ssps. *caeruleus* (N) and *ogliastrae* (C and S): very common (millions of pairs) and increasing; widespread, but local in much of SE and treeless areas. Ssp. *balearicus*: Mallorca (Serra de Tramuntana and vicinity), uncommon.

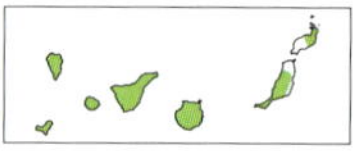

Canary Blue Tit *Cyanistes teneriffae* (herrerillo canario). ~11 cm. Recently split from Common Blue Tit. Cap and other head markings bluish-black, not blue: upperparts variably blue or greenish. Bill longer and finer ✦ In Canary Is, in laurel forests, pinewoods, tamarisks, palm groves, croplands ✦ Sedentary ✦ Ssps. *ombriosus* (El Hierro), *palmensis* (La Palma), *teneriffae* (La Gomera, Tenerife and Gran Canaria) and *degener* (Lanzarote and Fuerteventura): all common except more local *degener*; *ombriosus*, *palmensis* and *degener* considered threatened; **EN**. Ssp. *ultramarinus* (NW Africa): very common resident in Ceuta.

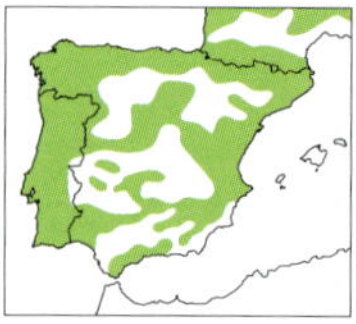

European Crested Tit *Lophophanes cristatus* (herrerillo capuchino europeo). 11 cm. Crest and facial pattern unique. Small, brownish. Call distinctive, although recalls Long-tailed Tit: a sober *cherr* or *cheer*, often preceded by a series of weak *tsi* or *sri* calls. Sexes similar; juvenile has shorter crest and weaker face pattern ✦ Forests, mainly coniferous (pines, spruces, junipers) but also in mature pedunculate, sessile, cork and holm oak woods and beechwoods. At all elevations. Uses or excavates holes in trunks and stumps. Occupies nestboxes ✦ Sedentary, with dispersion and autumn/winter wandering ✦ Ssps. *mitratus* (N and E) and *weigoldi* (SW): common (around 1 million pairs), perhaps increasing, although still quite concentrated in pinewood regions and mountains.

Great Tit
♂
♀
Common Blue Tit
♂ ad.
ssp. *teneriffae*
Canary Blue Tit
♂ ad.
ssp. *palmensis*
ssp.
ombriosus
♂ ad.
♂ ad.
ssp. *degener*
European Crested Tit

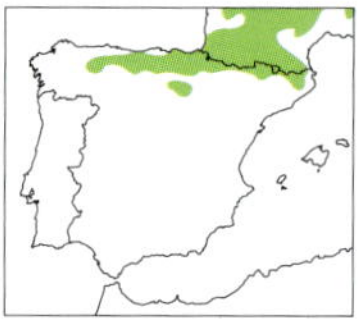

Marsh Tit *Poecile palustris* (carbonero palustre). 12 cm. Only in N. Cap and bib black, as in Coal Tit, but bib smaller and lacks white nape patch. Plain brown above, lacks grey. Sexes similar; juvenile cap not glossy. Calls include a diagnostic *pit–cháa* often followed by a scolding *checheche…* Song is an often repeated monotonous note, e.g. *schip–schip–schip…*) ✦ Woods, chiefly beech and pedunculate/sessile oaks, generally in mountains. Uncommon in riparian woods or countryside copses ✦ Sedentary ✦ Scarce (tens of thousands) and limited to the humid northern fringe and mountains of northern Iberian range; locally common, scarcer towards the west.

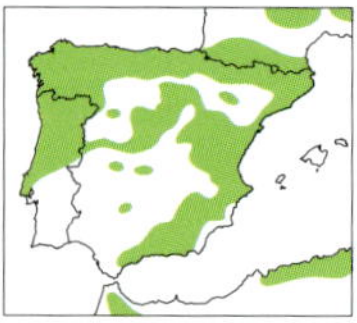

Coal Tit *Periparus ater* (carbonero garrapinos). 11 cm. Cap black, as in Great and Marsh Tits, but has obvious white nape patch. Lacks yellow and belly stripe. Bib very large, two white wingbars, grey above, flanks brownish. Bill long, fine. Juvenile greenish above, face pattern dirtier. Calls thin and sharp: *tsui*, *chi–chi–chi*, *si–si–si* (kinglet type); repetitive songs recall Great Tit but always bisyllabic: *pichú–pichú… tícha–tícha…* ✦ Woodlands, preferably pinewoods and including young plantations, at any altitude. Visits parks and gardens. Nests on ground in holes in banks or between tree roots; often in nestboxes ✦ Sedentary, with winter nomadism and arrivals of some European birds ✦ Ssps. *ater* (Pyrenees) and *vieirae* (elsewhere): very common (several million pairs); spreading, but largely restricted to montane pinewoods in C and S. Vagrant to Balearic Is.

PENDULINE TITS (Passeriformes, Remizidae) – Some ten species of small, generally insectivorous and woodland birds. Allied to tits. Nests highly elaborate. Eurasia, Africa and N America.

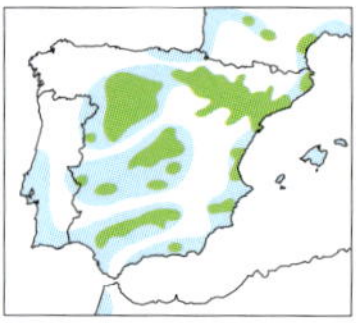

Eurasian Penduline-tit *Remiz pendulinus* (pájaro moscón europeo). 11 cm. Striking black mask on grey head; rufous back. Tail longish. Bill conical, sharply pointed. Buffish below. Forages in riparian or marshland vegetation, frequently giving a thin, whistle, downslurred *tsii*. Song unremarkable. Female and juvenile duller. Juvenile lacks mask ✦ Hangs pouch-like nest from poplars and willows alongside rivers, lakes and canals. Visits reedbeds and reedmace heads, especially in winter ✦ Spanish birds sedentary or dispersive; some numbers of overwintering (Sept/Oct to Feb/Mar) C European birds ✦ Uncommon (a few tens of thousands); frequent in Ebro valley; uncommon, but increasing, elsewhere. More widespread in winter. Scarce in Balearics.

PARROTBILLS (Passeriformes, Panuridae) – Only one species, in Europa and Asia.

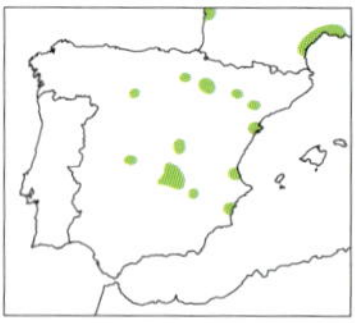

Bearded Parrotbill *Panurus biarmicus* (bigotudo). ~13 cm. Reedbeds. Cinnamon-coloured, short-winged and very long-tailed. Tit-like in appearance and behaviour. Often in small flocks, giving frequent *ping* and *chirr* calls. Moustached adult male, with black undertail coverts, unmistakable; juveniles of both sexes yellowish-buff, with black back and tailsides; juvenile male has black lores and yellow bill. Flight appears weak, with whirring wingbeats and occasionally fanned and twisted tail. Nests in loose colonies ✦ Marshland vegetation (reedbeds, canebrakes, reedmace) ✦ Resident but irruptive ✦ Scarce breeder (hundreds of pairs, decreasing), confined to the eastern Mediterranean coastlands (Albufera, El Hondo), La Mancha lagoons and a very few other inland sites. **NT**

Coal Tit
Marsh Tit
Eurasian Penduline-tit
♂ juv.
♀ ad.
♂ ad.
Bearded Parrotbill

LARKS (Passeriformes, Alaudidae) – Cryptic, terrestrial birds of open country. Nest on the ground. Run freely. Eat insects and seeds. Most species in Africa and Eurasia.

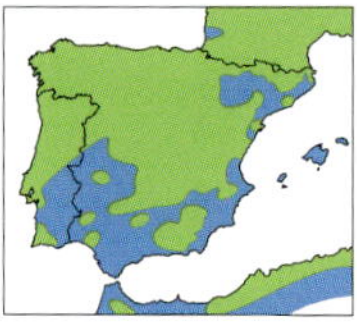

Eurasian Skylark *Alauda arvensis* (alondra común). ~18 cm. White outer feathers on longish tail and white trailing edge to wing evident in flight. Small erectile crest. Bill short and stout. Juvenile virtually lacks crest; shows pale-edged dark feathers on mantle and scapulars. Call: a pleasant, liquid *chirrup*. Song excellent: given largely in high song flight, an incessant and highly musical, varied, stream of gurgling, chirruping and whistling notes. Highly gregarious in winter ✦ Pastures, herbaceous crops, irrigated land, sparse scrub, etc., at varied altitudes. Common on stubbles and fallows in winter ✦ Spanish birds resident; European birds (ssps. *arvensis* and *scotica*, latter rare) in winter and on passage (Oct–Nov and Feb–Mar) ✦ Ssps. *cantarella* (NE), *guillelmi* (NW) and *sierrae* (S): common breeders (a few million); abundant in N half but scarcer in S, where occupies saltflats and high montane grasslands. Extremely abundant in winter (millions) and widespread, including Balearics. Scarce in Canary Is.

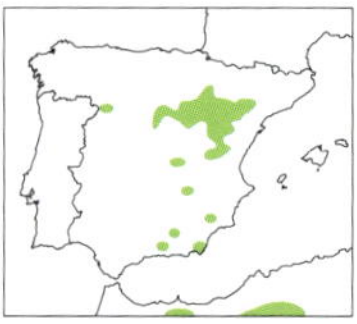

Dupont's Lark *Chersophilus duponti* (alondra ricotí). ~18 cm. Scarce and secretive, runs and hides between tussocks. Call, a modulated human-like whistle, prolonged at the end: *hu–heee*. Resembles a crestless Crested Lark but bill longer, more slender, and decurved; shows pale feather edging above; supercilium indistinct. Sings from ground or in flight, chiefly at dawn and dusk, often at night, mixing other notes into typical call. Poses very erect ✦ Sparse dwarf scrublands, often halophytic (thymes, brooms, false esparto grass, wormwood, etc.), typically level, with pebbly or gypsum soils, at various altitudes ✦ Resident, with short dispersive and cold weather movements ✦ Very scarce (a few thousand) and declining. Very local, with concentrations especially on the *páramos* of the Iberian range and in arid sectors of the Ebro valley. **EN**

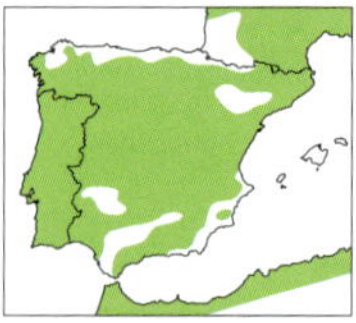

Woodlark *Lullula arborea* (alondra totovía). ~15 cm. A lark with distinctive head markings: long white supercilia that meet on nape and rufous-brown face; prominent black and white mark at wing bend; tail short, white-tipped; bill slender, crown feathers erectile. Call note: a sweet *tloo-eee*. Song highly musical, with clearly defined elements and typical repetitions, often accelerating and then falling in pitch: *liu–liu–liu–liu…* Sings in flight or from perch (on branch or wire), sometimes at night. Not very gregarious ✦ Dry pastures and sparse scrubland, with some trees (junipers, oaks, pines, almonds), dehesas, woodland rides and clearings. Ranges to some altitude in mountains ✦ Resident, with altitudinal movements; some Central European birds over-winter ✦ Ssps. *arborea* (N) and *pallida* (elsewhere): common (hundreds of thousands of pairs); uncommon in Biscay coastlands and very arid or deforested regions. Vagrant to Balearic Is.

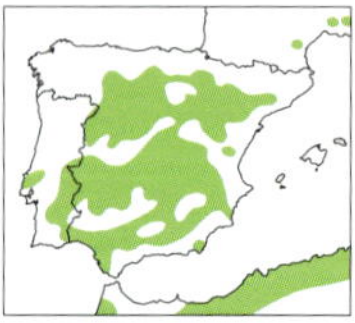

Calandra Lark *Melanocorypha calandra* (calandria común). ~19 cm. Robust, large-billed. Black neckside patches obvious when singing, poorly evident in fresh autumn plumage. Wings broad with white trailing edge, blackish below. Song less structured and musical than Skylark, with varied loud phrases, frequent mimicry and continuous insertion of harsh notes (*cherrrr*). Forms large winter flocks; nests in loose groups ✦ Emergent crops, fallows, dry pastures, saltflats, esparto grassland. Avoids scrub and irrigated land ✦ Largely sedentary but may wander ✦ Common (a few million), perhaps declining. Abundant in cereal croplands of the interior; absent from the humid north, mountains and much of the Mediterranean littoral. Occasional in the Balearic and Canary Is.

Eurasian Skylark
Dupont's Lark
Woodlark
Calandra Lark

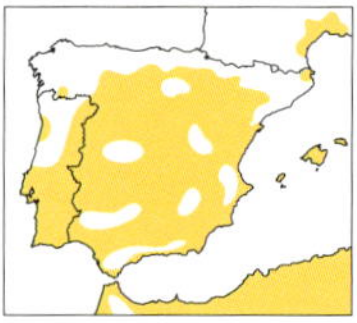

Greater Short-toed Lark *Calandrella brachydactyla* (terrera común). ~15 cm. Small sand-coloured lark, largely unstreaked below. Dark neck side patches most obvious when singing, poorly apparent in fresh autumn plumage, as in Calandra Lark. Bill short, conical. Streaked crown rufous, erectile; obvious supercilium and eye-ring; tertials reach tail tip. In flight, blackish tail shows brown centre and white sides. Juvenile shows pale feather edging above. Calls sparrow-like: *chirp* or *chip*. Song distinctive, given in flight, starting with a variable introduction at take-off and then comprising identical phrases repeated with brief pauses, in synchrony with flight undulations: *tii–tii–ti–ti–ti–tu–tu–tu*. Migrates in flocks. Often nests in loose groups ✦ Level or gently sloping dry areas, with some bare ground (fallows, saltflats, dwarf-shrub steppes, vineyards) ✦ Summer visitor (late Mar/Apr–Sept/early Oct) ✦ Common (hundreds of thousands of pairs) but rapidly declining. Frequent throughout Mediterranean region (Andalucía, La Mancha, Ebro valley), scarcer towards N and NE. Rare migrant in Canary Is. **VU**

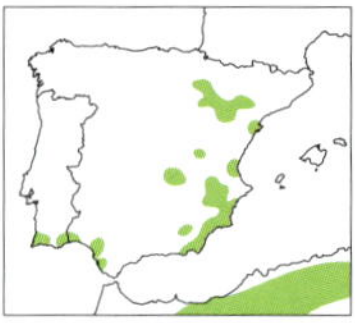

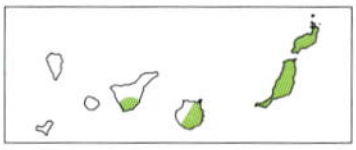

Lesser Short-toed Lark *Alaudala rufescens* (terrera marismeña). ~14 cm. Greyer than Greater Short-toed and strongly streaked on breast and flanks. Bill shorter (linnet-like); supercilium and eye-ring poorly marked. Primaries project >1 cm beyond tertials. Calls different, more buzzing: *prrrit*, *chirrrri*… Song more varied and musical; recalls Skylark but sharper and includes harsh call notes and mimicry. Highly gregarious year-round ✦ Natural dwarf-shrub 'steppes', with much bare ground, often on saline or gypsum soils (false esparto, white wormwood, saltwort, glasswort, esparto) ✦ Sedentary, with nomadism ✦ Ssp. *apetzii*: some two hundred thousand pairs, decling; chiefly in Ebro valley, the arid SE and Andalucian marismas. Ssp. *rufescens*: Tenerife, practically extinct, **CR**. Ssp. *polatezki*: E Canary Is, widespread and common, **EN**. Vagrant to Balearic Is.

Crested Lark *Galerida cristata* (cogujada común). ~18 cm. Common. Distinctive crest; shows no white in flight. Plump, long-billed, warm brown; tail dark brown with rufous outer feathers; underwings rusty brown (compare Thekla Lark). Juvenile has short crest and white spotting on cap and back. Distinctive call: *tu–tu–tíu*. Song typical lark type but relatively weak, with well-spaced notes; often sings from the ground. Unsociable ✦ Open farmlands, both dry and irrigated; often on outskirts of villages and along roadsides ✦ Sedentary; sparse winter immigration ✦ Ssps. *cristata* (N) and *pallida* (elsewhere): very common, particularly in C and S; absent in N and mountains (hundreds of thousands or a few million pairs), declining. Ssp. *kleinschmidti*: Ceuta and Melilla. Possible vagrant to Canary Is.

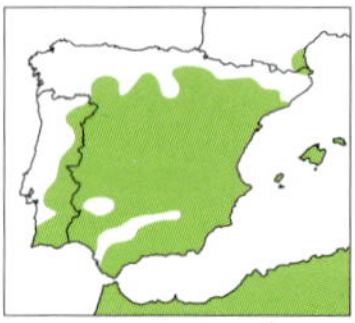

Thekla Lark *Galerida theklae* (cogujada montesina). ~17 cm. Difficult to separate from Crested Lark but slimmer, greyer; bill shorter and straight; crest shorter and more compact; breast streaking coarser and more defined; underwing greyish, not rufous. Call louder and more energetic, with notes in crescendo: *tu–tU–TÍU*. Song more varied and cheerful, in flight or often from a perch (on stone, shrub, wire; Crested Lark seldom uses high perches) ✦ Fallows, tall scrub, esparto grasslands, dry pastures, bushy hillsides…, often on stony sloping ground; also in mountains. Generally avoids woods and croplands ✦ Sedentary ✦ Ssp. *theklae*: very common (many hundreds of thousands or a few million) abundant in SE and dry mountains; progressively rarer towards N; very common in Balearics. Ssp. *erlangeri*: Ceuta and Melilla.

Greater Short-toed Lark
Lesser Short-toed Lark
Crested Lark
Thekla Lark

CISTICOLAS AND ALLIES (Passeriformes, Cisticolidae) – Small greenish or brownish insectivores, with medium-long or long tails. Habitats varied. Tend to build purse-shaped nests. Most in Africa and Asia; one species in Europe.

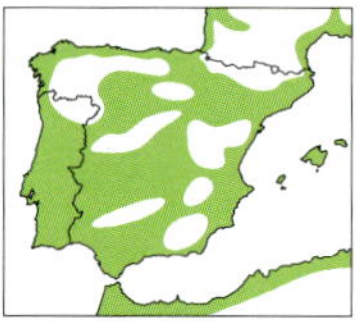

Zitting Cisticola *Cisticola juncidis* (cistícola buitrón). ~10 cm. Tiny. Calls *tsip…tsip…tsip…* monotonously in flight, climbing further with each note. Wings short and rounded, tail graduated, often drooped and half-fanned in flight. Perches on vertical stems, jerking tail. Mantle clearly streaked; pale face looks plain; throat whitish; rectrices tipped black and white ✦ Dense herbaceous vegetation, often near wetlands: marsh and lake fringes, sedgy pastures, ricefields and other croplands, chiefly irrigated ones. Enclosed nest built near ground. Cold-sensitive, avoiding elevations above 1,000 m ✦ Sedentary and dispersive ✦ Ssp. *cisticola*: quite common (many tens of thousands), increasing; abundant in Mediterranean coastlands, Andalucía and part of Extremadura, but uncommon on N Meseta and absent from forested and montane areas. Possibly vagrant to Canary Is.

REED WARBLERS AND ALLIES (Passeriformes, Acrocephalidae) – Brownish or olive coloured insectivores, often streaked, with fine, straight bills. Formerly included in warbler family (Sylviidae). Many frequent marshland vegetation. Eurasia, Africa and Oceania.

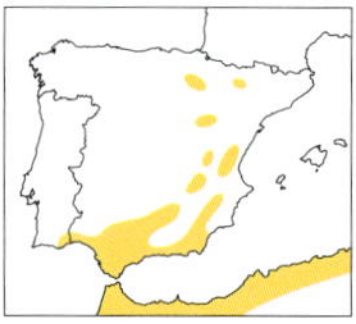

Isabelline Warbler *Iduna opaca* (zarcero bereber). ~13 cm. Also known as Western Olivaceous Warbler. In E and S. Rather nondescript and identification needs care. Lacks greenish or yellow tones, unlike Melodious Warbler (below). Coloration recalls Garden Warbler (p. 180) but bill longer with pale lower mandible; head pattern more defined, primary projection shorter and tail sides pale. Also recalls Common Reed-warbler (p. 170) but tail square-ended and undertail coverts short. Song also like Reed-warbler, grating and repetitive, but more rapid with frequent changes of pitch. Call: *chac*. Often nests in loose groupings ✦ Riparian growth (tamarisks, willows, oleanders, canes), also poplar stands, orchards, parks; typically in hot, dry areas ✦ Summer visitor (late Apr/May–Aug/Sept), trans-Saharan migrant ✦ Scarce (a few thousand pairs); mainly in S and SE, very local elsewhere. Rare migrant in Balearics and Canary Is. **NT**

Melodious Warbler *Hippolais polyglotta* (zarcero común). ~13 cm. Common. Largely olive above and yellow below, although yellow very pale in some leading to possible misidentification as Isabelline or Garden Warbler. Willow Warbler is smaller, fine-billed, with pale legs and obvious supercilia and eye-stripes. Song has two parts: an introductory slow repetition of one note followed by a rapid and erratic babble. Calls sparrow-like: *trr*, *chek–chek* ✦ Tall shrubbery: bramble clumps, thorn hedges, tamarisks, orchards ✦ Trans-Saharan migrant, passage (~Apr–May and Aug–Oct) ✦ Common (hundreds of thousands or a few million) and widespread. Rare in humid mountains. Uncommon migrant in Balearics. Scarce migrant in Canary Is.

Icterine Warbler *Hippolais icterina* (zarcero icterino). ~13 cm. Rare in most regions. Resembles Melodious Warbler but has longer primary projection; about equal to tertial length; half this in Melodious. Pale edges to secondaries and tertials produce a pale wing panel, less evident in worn adults in autumn. Legs blue-grey; usually brown in Melodious ✦ Shrubbery, copses ✦ Migrant (May and Aug–Sept). Nests eastwards from NE France ✦ Regular in Balearics, especially in spring; scarce or uncommon in littoral E and NE; occasional elsewhere. Vagrant to Canary Is.

Zitting Cisticola
Isabelline Warbler
Melodious Warbler
Icterine Warbler

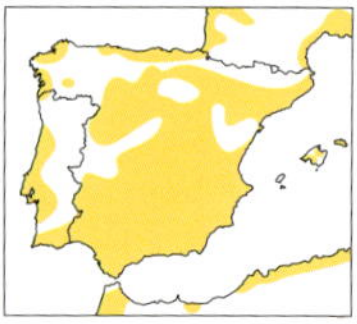

European Reed-warbler *Acrocephalus scirpaceus* (carricero común). ~13 cm. Common. Resembles Great Reed Warbler but much smaller. Also confusable with rare species such as the **Marsh Warbler** *Acrocephalus palustris* (carricero políglota) (p. 235), or with the Melodious Warbler. The reed-warblers are warm brown, flat-crowned (forehead-less); tail rounded, unpatterned; undertail coverts very long. Juveniles brighter-toned, supercilium less evident. Song is long series of harsh notes, repeating notes 2–3 times: *chirrí–chirrí, pia–pia–pia, tret–tret…* ✦ Mainly in reedbeds, at lakes, marshes, slow streams, irrigation channels ✦ Trans-Saharan migrant, with passage of European birds (Apr–May and Aug–Nov) ✦ Breeds quite commonly (many tens or some hundreds of thousands), increasing and widespread; the commonest reedbed passerine. Very common on passage. Rare and irregular in Canary Is.

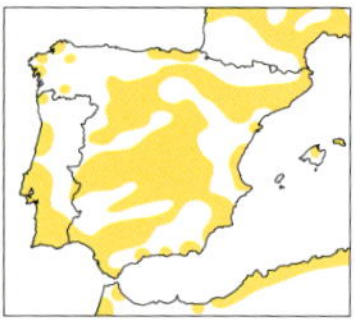

Great Reed-warbler *Acrocephalus arundinaceus* (carricero tordal). ~19 cm. Starling-size. Bill robust; broad supercilium; crown slightly peaked, often raised when singing. Song similar to Common Reed-warbler but louder, harsh, deliberate and lower pitched, with characteristic repetitions *carra–carra, gurk–gurk…* Juvenile more rufous. Shows itself readily ✦ Reedbeds, sometimes also reedmace and cane clumps; also along riverbanks ✦ Trans-Saharan migrant; passage inconspicuous (Apr–May and Jul–Oct) ✦ Relatively common (many thousands of pairs), although less widespread than Common Reed-warbler; very frequent in E and S, scarce in N coastlands and absent from elevated regions. Vagrant to Canary Is.

Sedge Warbler *Acrocephalus schoenobaenus* (carricerín común). ~13 cm. Regular migrant. Dark cap has indistinct suggestion of pale median stripe; eyestripe long, broad, creamy; dark loral line. Olive-brown above, faintly streaked (not boldly marked as in Aquatic Warbler); rump plain, reddish brown; flanks unstreaked. Worn autumn adults, and juveniles, show breast-side spotting ✦ Reedbeds ✦ Passage (Mar–May and Jul–Oct) of birds from Britain and elsewhere in W Europe that winter in Africa (nests from SW France onwards) ✦ A few pairs have nested on rare occasions in Basque Country. Uncommon but widespread on passage. Scarce in Balearics. Occasional in Canary Is.

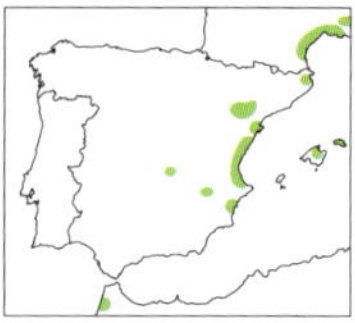

Moustached Warbler *Acrocephalus melanopogon* (carricerín real). ~13 cm. Cap blackish; broad, widening, white supercilium ends abruptly; dark ear-coverts and moustachial streak. Dark chestnut above, streaked; most rufous in spring. Primary projection short; <1/3 tertial length. Rounded tail, often raised. Legs dark. Song more varied and gurgling than reed-warblers, less harsh, including a woodlark-like *liu–liu–liu–liu…* Forages low down, often on reed mats ✦ Reedbeds, reedmace and sedge beds, etc., mainly near the water ✦ Sedentary, with dispersion and wintering of French-origin birds (Oct–Feb) ✦ Scarce (~a thousand pairs), declining, locally common (Alicante and Valencia wetlands, Albufera de Mallorca). Very scarce inland. **VU**

Aquatic Warbler *Acrocephalus paludicola* (carricerín cejudo). ~13 cm. Scarce. Highly streaked, yellowish, shows distinct pale yellow median crown stripe as well as supercilia. Two yellowish lines flank mantle. Rump streaked. Lacks loral line. Legs pinkish. Often forages on ground ✦ Reedbeds and other marsh vegetation ✦ On passage (~Mar–May and Aug–Oct). Nests eastwards from Poland but has winter quarters in W Africa ✦ Very scarce but regular, most frequent in late summer/early autumn; most often detected by ringers, both at coastal sites and inland (laguna de La Nava). Vagrant to Balearic and Canary Is. **VU**

European Reed-warbler
Great
Reed-warbler
Sedge Warbler
Moustached
Warbler
Aquatic Warbler

GRASSHOPPER-WARBLERS (Passeriformes, Locustellidae) – Allied to reed-warblers. Typically brown, streaked. Inhabit dense vegetation. Old World.

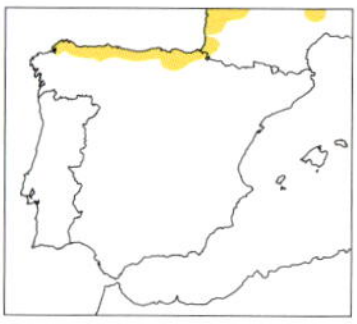

Western Grasshopper-warbler *Locustella naevia* (buscarla pintoja). ~13 cm. Nest in N. Song an insect-like, mechanical, high-pitched reeling, varying in loudness as bird turns head; long-lasting and far-reaching. Highy skulking, creeping through undergrowth like mouse. Brownish, streaked above. Tail long, graduated ✦ Dense grassy scrubland in heathland, young plantations, streamsides, woodland clearings ✦ Trans-Saharan migrant, passage (~Mar–May and Aug–Oct) ✦ Scarce breeder (a few thousand), only at low levels in the humid Biscay coastlands. Scarce on passage but more frequent on E coast. Occasional in Canary Is.

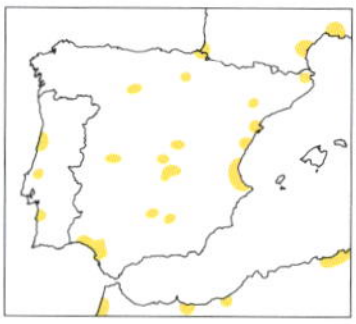

Savi's Warbler *Locustella luscinioides* (buscarla unicolor). ~14 cm. Reedbeds. Insect-type reeling similar to Grasshopper Warbler but faster, lower-toned, also wavers in volume. Unmarked, as reed-warblers, but differs in being dark rufous-brown above, darker on breast and flanks and especially in rounded tail; often raised and jerked energetically. Hard to see ✦ Extensive reedbeds and other large wetlands ✦ In summer and on passage (late Mar–May and Aug–Oct), trans-Saharan migrant ✦ Scarce nester (hundreds of pairs), apparently spreading. Most frequent near coasts (Marismas, Valencia, Delta) but also at inland wetlands (esp. in La Mancha). Rare in the Balearics; may breed Albufera de Mallorca. **NT**

SWALLOWS (Passeriformes, Hirundinidae) – Specialists in capturing insects in flight. Very long-winged, short-legged, with tiny beaks and large gapes. Flight often slow; highly manoeuvrable. Attach mud nests to rocks and walls. Many highly migratory. Worldwide.

Barn Swallow *Hirundo rustica* (golondrina común). ~18 cm. Only swallow that is uniformly blue-black above. External rectrices elongated, in males especially. Whitish below, reddish in males in fresh plumage. Bib and forehead red; collar blue-black; tail shows white ovals. Juvenile not glossy; bib and forehead buffish-red; collar less distinct; external rectrices short. Weak song, often given in flight, an agreeable twittering with occasional rattling. Often hunts low down. Perches on wires and fences. Gregarious; large numbers roost in reedbeds ✦ Open countryside and town outskirts. Nests in stables, under eaves, in porches, in ruins ✦ In summer and on passage (~Feb–May and Aug–Oct), trans-Saharan migrant ✦ Very common (hundreds of thousands or a few million pairs, declining) and widespread, although scarce in wooded areas or above 1,500 m. Some present in winter in S. Common migrant in Canary Is.

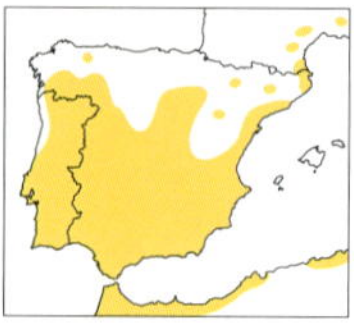

Red-rumped Swallow *Cecropis daurica* (golondrina dáurica). ~17 cm. Resembles Barn Swallow but rump is reddish, whiter towards rear; lacks pectoral band; tail streamers thicker, tail lacks white ovals; vent black. Juvenile duller, lacks tail streamers. Less vocal than Barn Swallow. Usually met as isolated pairs ✦ More frequent than Barn Swallow in hilly or wooded regions; half-bottle nest built on roofs of small caves or under bridges, less often on buildings ✦ Trans-Saharan migrant (mainly Mar/May–Sept/Oct) ✦ Ssp. *rufula*: scarce (many tens or hundreds of thousands), breeding range spreading slowly from Andalucía; very common in parts of SW but local in N half. Scarce migrant in Balearics. Regular in Canary Is in spring.

Western Grasshopper-warbler
Savi's Warbler
Barn Swallow
Red-rumped Swallow

House Martin *Delichon urbicum* (avión común). ~14 cm. The swallow that commonly nests under eaves and balconies. Unmistakable black and white plumage; glossy blue-black above, with striking white rump, and pure white below (swifts with white-rumps, see p. 42, have black underparts). Juvenile duller, with whitish-edged remiges. Call: *chirrp*. Song shorter and less varied than Barn Swallow, with rasping notes. Highly gregarious; often nests in colonies of dozens or hundreds of pairs. Perches on wires. Tends to feed higher up than swallows ✦ Nests on buildings, bridges, dams, etc., also on cliffs, at altitudes up to 2,500 m. Often in town centres ✦ Trans-Saharan migrant (mainly Feb–Apr and Sept–Nov), some present in S in winter ✦ Very common (some millions) and widespread. Does not nest in Canary Is, but common there on passage.

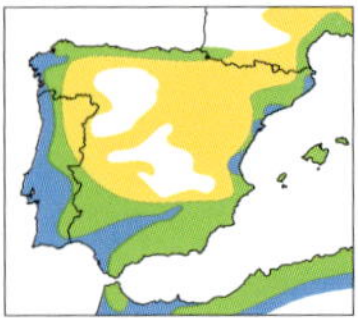

Crag Martin *Ptyonoprogne rupestris* (avión roquero). ~15 cm. Brownish, lacks pectoral band. Largish and compact, tail forked, with diagnostic white spots. Dusky overall though paler below; in flight shows contrasting dark underwing coverts, axillaries and vent. Throat speckled. Juvenile has rufous-brown feather tips above. Not very vocal; calls weakish: *trrrit*. Often flies close to rock faces, with frequent changes in direction after 'stopping' in fight. Groups often sun themselves on ledges and rooftops. Nests alone or in loose colonies ✦ Broken terrain at variable altitudes. Nests similar to Barn Swallow; on rock faces and also on bridges, dams, tunnel roofs; sometimes on houses ✦ Migratory; mainly summer visitor inland (Feb/Mar–Sept/Nov), wintering in warm regions, winterers including birds from W European countries ✦ Quite common (tens of thousands of pairs), absent only from very flat regions. Large numbers winter in coastal regions and in S half. Vagrant to Canary Is.

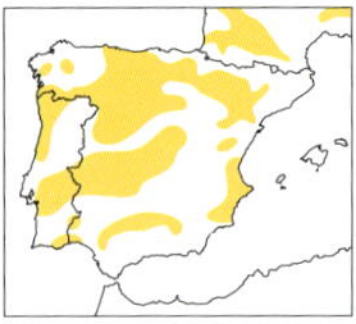

Sand Martin *Riparia riparia* (avión zapador). ~12 cm. Brown, small, has pectoral band. Tail forked. Two-toned; earth-brown above, white below with dark underwings. Juvenile shows pale feather edging above. Calls brief, short: *churr*, *chrrip*. Often skims water bodies. Very gregarious; often in colonies of hundreds of pairs ✦ Nests are tunnels in sand banks near water or often within gravel and sandpits. Generally in broad valleys at low elevations ✦ Trans-Saharan migrant, with strong passage of European birds (late Feb–May and Aug–Oct); some occur in S in winter ✦ Quite common (many tens of thousands or a few hundred thousand), most frequent within the Duero and Ebro basins. Common migrant in Balearics. Regular in Canary Is.

BULBULS (Passeriformes, Pycnonotidae) – Mainly forest birds. Frugivorous and insectivorous. Plumages typically dull, often with white or coloured areas on head and undertail coverts. Mainly tropical regions of Africa and Asia.

Common Bulbul *Pycnonotus barbatus* (bulbul naranjero). ~20 cm. Slender and long-tailed, plumage dull. Head and tail somewhat blacker and undertail coverts whitish. Starling-sized. Sexes similar. Often raises crown feathers. Wings broad and rounded. Forages in trees and shrubs, flight looks weak. Gregarious. Cheerful and confiding. Very garrulous, with various calls; sings year-round; song: a series of short, diverse phrases, including a fluting *tee–tuee* or *tee–tu–teeú* ✦ In Africa frequents *wadis*, bushy river courses, gardens, orchards, orange groves, palm clumps ✦ Resident, with post-nuptial dispersion ✦ Common at low levels in Morocco and Algeria, small numbers breed in Ceuta (Calamocarro and Las Bombas streams). Vagrant to S Andalucía, has bred Tarifa (Cádiz).

House Martin
Crag Martin
Sand Martin
Common Bulbul

LEAF-WARBLERS (Passeriformes, Phylloscopidae) – Small greenish or yellowish woodland birds, with fine bills and relatively long wings. Formerly included in Sylviidae (Old World warblers). Sexes similar. Insectivorous. Typically migratory. Eurasia and Africa.

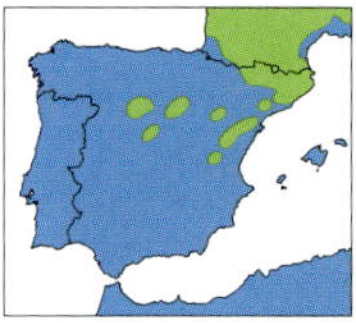

Common Chiffchaff *Phylloscopus collybita* (mosquitero común). ~11 cm. Year-round. Darker and more compact than Willow Warbler; wings shorter, primary projection < half tertial length; bill and legs blackish. Brownish-olive above, buffy or yellowish-white below. Supercilium shortish and indistinct; alula blackish. Call: a soft, rising *huit*. Song, a measured repetition of two notes, alternated irregularly: *chif–chaf–chif–chif–chaf…* ✦ Nests in damp woodlands and scrub, and in riparian woods. Very widespread in winter in scrub, parks, reedbeds, riversides, etc. ✦ Migratory within Europe, most wintering in the Mediterranean region. On passage Sept–Nov and Feb–Apr ✦ Ssp. *collybita*: common breeder (tens or hundreds of thousands of pairs) in Pyrenees and Catalonia and also in the riparian woodlands of the Duero basin; much more local elsewhere but distribution not fully known; abundant (some millions) on passage and in winter, especially in warmer coastal regions. Abundant in winter in Balearics. Scarce in Canary Is. Ssp. *abietinus*: Rare migrant (from Scandinavia and Russia). Ssp. *tristis* (Siberia): very rare, has wintered.

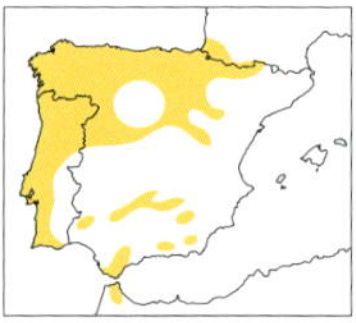

Iberian Chiffchaff *Phylloscopus ibericus* (mosquitero ibérico). ~11 cm. Until recently treated as a subspecies of Common Chiffchaff. Similar but more brightly coloured, longer-winged, supercilium more obvious, bill and legs brownish; often indistinguishable in the field. Song distinctive however, a single short phrase repeated several times and accelerated towards the send: *chif–chif–chif–chif–swit swit swit* (mixed singers, possiby hybrids, occur). Call different: a plaintive, downslurred *sweeoo* ✦ Humid woodlands and scrub, often along streams and rivers ✦ Summer visitor, trans-Saharan migrant (late Feb/Apr to Aug/Oct) ✦ Quite common breeder (hundreds of thousands of pairs), very frequent in Galicia and the Biscay coastlands; much less common on N Meseta; isolated populations in rainier parts of the S (Sierra Morena, Montes de Cádiz).

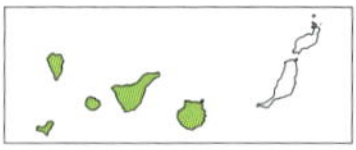

Canary Chiffchaff *Phylloscopus canariensis* (mosquitero canario). ~10 cm. Recently separated from Common Chiffchaff; darker, smaller and rounder-winged. Song similar but more rapid and explosive. Call often disyllabic ✦ All habitats other than the most arid, including woodlands, scrub, croplands and gardens ✦ Sedentary. Nests Jan–Jun ✦ Abundant (tens of thousands of pairs) and very widespread, although absent in Lanzarote and Fuerteventura (where local race *exsul*, now extinct). **DD**

Willow Warbler *Phylloscopus trochilus* (mosquitero musical). ~11 cm. Migrant. Neater than Common Chiffchaff, brighter-coloured, bill and legs pale, facial markings more distinct; primary projection nearly as long as tertials. Very yellow below in autumn, particularly juveniles. Call longer and disyllabic: *hoo–eet*. Song, a cadence of musical notes, rapid at first, then slowing to a brief flourish ✦ Nests in open deciduous woodlands, often on shady bushy slopes near streams ✦ Tran-Saharan migrant, passage (~late Mar–early May and Aug–Oct) (ssp. *trochilus* and *acredula*) ✦ Only a few tens of pairs nest, very locally in the Cantabrian range (mainly in Asturias) and Basque Country; **NT**. Abundant on passage, when also common in Balearics and regular at least in the eastern Canary Is.

Common Chiffchaff
Iberian Chiffchaff
Canary Chiffchaff
Willow Warbler

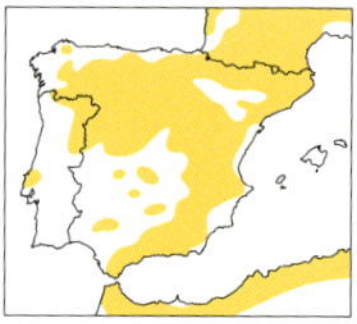

Bonelli's Warbler *Rhadina bonelli* (mosquitero papialbo). ~11 cm. Summer visitor. Very white below and pale-headed; dark eyes stand out against pale face and lores; supercilium indistinct. Wings and tail edged green; rump yellowish. Call disyllabic: *chew–ee*. Song a short trill on one note; may recall Cirl Bunting ✦ Open, not particularly humid, woodlands, mainly of Pyrenean or Portuguese Oaks; also Holm Oaks, pines, junipers, etc. ✦ Trans-Saharan migrant, passage in Apr–May and Aug–Sept ✦ Common (a few million) and widespread, although scarce in Biscay coastlands and Galicia and tends towards montane regions, especially in S. Scarce migrant in Balearic and Canary Is.

Wood Warbler *Rhadina sibilatrix* (mosquitero silbador). ~12 cm. Uncommon migrant. Face, throat and upper breast bright lemon yellow, contrasting with very white underparts and green upperparts. Wings very long; tail short; remiges edged yellow, blackish tertials edged pale; legs yellowish. Call: a mournful *piu*. Song: repeated *piu* notes followed by a long, metallic, accelerating trill ✦ In W Europe nests in shady woodlands with little understorey, often beechwoods or oakwoods ✦ Trans-Saharan migrant, passage (~Apr–May and Aug–Sept). Nests commonly from France onwards but migrates SE ✦ A very few pairs nest, perhaps irregularly, in the Pyrenees and Cantabrian Range. Scarce migrant, most frequent in spring and in Catalonia and the Balearics. Scarce but regular in Canary Is.

BUSH-WARBLERS (Passeriformes, Scotocercidae) – A small family of insectivores with long, straight bills and rounded wings and tails. Plumages often brownish. Eurasia and NW Africa.

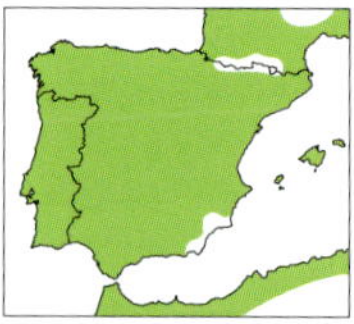

Cetti's Warbler *Cettia cetti* (cetia ruiseñor). ~13 cm. Dense vegetation by water. Small, brown, moves discreetly in undergrowth. Explosive song diagnostic: *chip…chu-eep…coochee, coochee, coochee* (or 'cet–tí'), year-round; calls less distinctive: *chip*, *huit*, *chi–chi–chi–chi–chi*. Considerably smaller than Nightingale (p. 192) and less rufous on rump and tail. Wings short and rounded; supercilium greyish, undertail coverts faintly pale-spotted ✦ Dense shrubbery along watercourses and around lakes, pools and damp hollows; also in reedbeds, especially in winter ✦ Sedentary or partly migratory, with some movement from interior to coasts ✦ Common (many tens or hundreds of thousands) and well distributed, although avoids mountainous or very dry regions. Nests commonly in Balearics, and in Ceuta.

BUSH-TITS (Passeriformes, Aegithalidae) – A dozen tit-like woodland species. Small-billed, with short wings and long, graduated tails. Build closed, purse-like nests. Eurasia and N and C America.

Long-tailed Tit *Aegithalos caudatus* (mito común). ~14 cm. Suggests a small tit with a tiny beak and a tail longer than its body. Usually in flocks, often giving thin *tsi–tsi–tsi* calls followed by rasping *tserr*, *tserr* (recalling Crested Tit). Greyish-brown, tinged pinkish below; head white with broad black side patches. Juvenile shorter-tailed, head pattern brown and white. Confiding, acrobatic ✦ Woodlands of all kinds, especially open shrubby ones; also riparian woods ✦ Sedentary, with short dispersal movements in winter ✦ Ssps. *taiti* (N) and *irbii* (C and S): common (hundreds of thousands of pairs); most frequent in N and C; in S chiefly in forested montane areas. Occasional in Balearics, has nested recently.

Bonelli's Warbler

Wood Warbler

Cetti's Warbler

Long-tailed Tit

SCRUB WARBLERS (Passeriformes, Sylviidae) – Small, fine-billed with rounded wings and variably long tails. Insectivorous and frugivorous. Eurasia and Africa.

Blackcap *Sylvia atricapilla* (curruca capirotada). ~14 cm. Cap black in male, chestnut-brown in female and juvenile. Differs from all other scrub warblers, except Garden Warbler, in absence of any white markings. Call: a sharp *tac* often repeated. Song begins as a gentle chatter soon turning into a series of well-spaced musical notes ✦ Woodland with understorey, riparian growth, humid open countryside with trees, orchards, parks. In winter frequents olive groves and Mediterranean woodland ✦ Partial migrant, with wintering and passage of European birds (~Mar–Apr and Sept–Nov) ✦ Ssps. *atricapilla* (N and C) and *heineken* (W, S and Ceuta): common (hundreds of thousands or a few million), increasing and spreading; very numerous in N and humid mountain regions, more local in S, where commonest along watercourses and shaded slopes. Abundant (millions) on passage and in winter, particularly in Mediterranean coastlands, Andalucía and Extremadura. Ssp. *pauluccii*: common in Balearics. Very common in C and W Canary Is (*heineken*), in woods, croplands and parks.

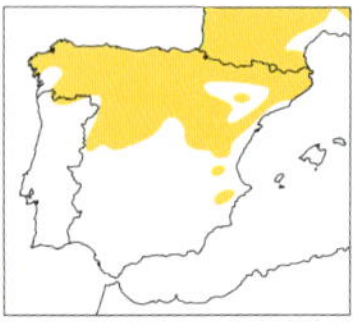

Garden Warbler *Sylvia borin* (curruca mosquitera). ~14 cm. Lacks distinctive fieldmarks. Head rounded; neck sides show greyish patches, eyes dark, with white eye-rings; bill relatively short and stout, paler at base; legs blue-grey. Sexes similar. Suggests a plump, cap-less Blackcap. Call: *check*, less harsh than Blackcap. Song recalls Blackcap but longer and more rapid. Skulking ✦ Shrub layer of humid woodlands, especially in clearings and woodland edges, willows and other riparian growth, dense hedges ✦ Trans-Saharan migrant with passage (Apr–May and Aug–Oct) of birds from C and N Europe ✦ Breeds quite commonly (hundreds of thousands) but virtually restricted to N and the mountains of C and E. Numerous on passage in Peninsula and Balearics. Regular migrant in Canary Is, particularly in E.

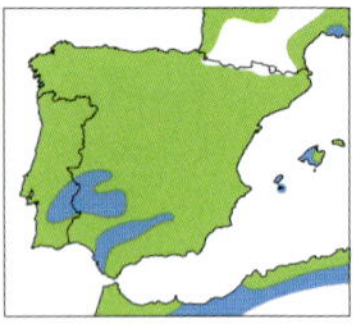

Dartford Warbler *Sylvia undata* (curruca rabilarga). ~13 cm. Small, dark, very long tail often cocked, wings short. Forages in shrubs, often moving short distances between them, flying low with tail jerking. Call: *tchurr* or *tee–chur*. Throat dark, streaked white. Adult wine-red below, brighter in male; eye-ring red, legs yellowish brown. Juvenile duller, breast variably dusky grey. Song, given from perch or in flight, resembes other small scrub warblers, a jumble of musical and harsh notes ✦ Scrublands throughout, of varying height and density at a wide range of altitudes ✦ Sedentary, with altitudinal movements ✦ Ssps. *undata* (NE), *dartfordiensis* (NW) and *toni* (C and S): very common (perhaps 2–3 million pairs) and widespread; scarce only in highly cultivated regions. Breeds Menorca and Mallorca (*undata*).

Balearic Warbler *Sylvia balearica* (curruca balear). ~13 cm. Endemic to Balearics. Only recently separated from **Marmora's Warbler** *S. sarda* (curruca sarda), vagrant (p. 236). Very similar to Dartford Warbler but adult grey below (tinted pinkish or brown but never wine-red or ochre), chin and throat whitish. Juvenile and 1st winter female have pale eye-ring; greyish head contrasts with brownish back. Call different: *tchek* or *tscherk* ✦ Diverse Mediterranean scrublands; avoids woodlands and cultivation ✦ Sedentary ✦ Ssp. *balearica*: very common (tens of thousands) and widespread in Mallorca, Cabrera, Ibiza and Formentera; extinct in Menorca, where perhaps displaced by Dartford Warbler.

Blackcap
♂
♀
Garden Warbler
♀
Dartford Warbler
♂
♂
♀
Balearic Warbler

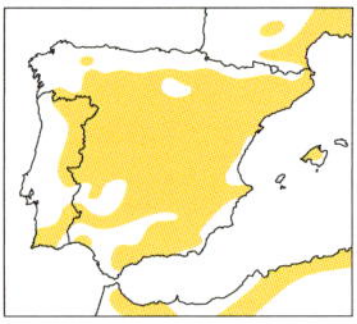

Subalpine Warbler *Sylvia cantillans* (curruca carrasqueña). ~12 cm. Adult male distinctive; grey above, brick-red below with large white moustachial stripe. Female browner above and creamy white below. Juvenile hard to separate from some other young scrub warblers; pale; brownish above, creamy below; orbital ring pale, barely a hint of moustachial stripe. Small, delicate; almost like a Dartford Warbler but tail shorter. Bill short, dark; tail dark, rounded, with white outer feathers. Call: *tec*, or *tec–tec–tec*... Song similar to other small scrub warblers, less harsh than Sardinian Warbler. Secretive ✦ Tall scrub, taller and denser than frequented by Dartford or Spectacled Warblers (holm and kermes oak scrub, scrubby dehesas, dry pinewoods, cistus scrub) ✦ Trans-Saharan migrant (~Mar/Apr–Aug/Oct) ✦ Ssp. *cantillans*: common (a few million) and increasing; frequent in Mediterranean regions with some forest or shrub cover; absent from Biscay coastlands and humid mountains. Common on passage in Balearics; scarce but regular migrant in Canary Is. Ssp. *moltonii*: breeds at least in Mallorca and Cabrera, scarce. Ssp. *albistriata*: vagrant to Balearic Is.

Lesser Whitethroat *Sylvia curruca* (curruca zarcerilla). ~13 cm. White-throated but lacks rufous on wings, unlike Greater Whitethroat and Spectacled Warbler; legs dark. Grey-brown above; grey head with darker cheek and lores, giving masked effect; rump grey; tail shortish, outer rectrices white. Call: a brief *chet* or *chec*. (See Subalpine and Sardinian Warblers) ✦ Typical of bushy woodland edges ✦ Trans-Saharan migrant, nests in Europe from France onwards, migrating SE (~Apr–May and Sept–Oct) ✦ Occasional, more frequent in autumn; most records in Balearics, Catalonia and Valencia.

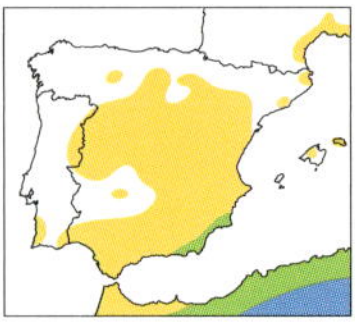

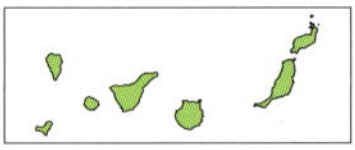

Spectacled Warbler *Sylvia conspicillata* (curruca tomillera). ~12 cm. Suggests a small, brightly-coloured Greater Whitethroat. Head looks large; orbital ring white, primary projection very short; tertials with well-defined dark centres. Moves briskly through very low vegetation. Call: *trrrrrr*, *trrrrrr*... Adult male has head blue-grey, lores dark; underparts pinkish, contrasting with white, greyish-centred, throat. Female similar; head less grey. Juvenile brownish above and pale, similar to juvenile Common Whitethroat. Often sings in flight; song a series of short whistled phrases ending in a rapid warble ✦ Scrubland, typical in shorter, sparser cover than other scrub warblers, often in arid regions, including mountainsides, salt flats and marsh fringes ✦ Summer visitor in N and inland (Mar–Oct) with possibly a few wintering locally in S and SE ✦ Ssp. *conspicillata*: rather uncommon (a few hundred thousand), most frequent in arid SE; rare on N Meseta, scarce in Balearics. Ssp. *orbitalis*: Canary Is, abundant and widespread; **DD**

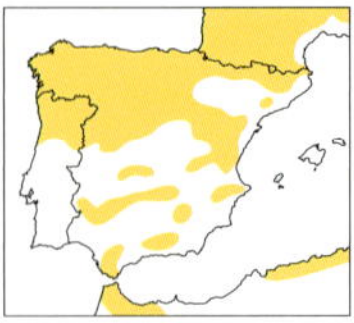

Greater Whitethroat *Sylvia communis* (curruca zarcera). ~14 cm. White throat and rufous wing panel striking; as Spectacled Warbler but larger and duller. Male has head grey when breeding; brown otherwise. Orbital ring white; iris reddish in adult; long tail has white external rectrices. Juvenile tawnier, throat less white. Calls various: *charr*, *wit–wit–wit–wit*... Song, given from perch or in flight, a brief chatter, longer and more varied in flight ✦ Shrubby woodland edges (in gorse or broom scrub), hedges and field margins, montane scrublands ✦ Trans-Saharan migrant, with passage (late Mar–May and Aug–Oct) of birds from C and N Europe ✦ Fairly common breeder (some hundreds of thousands), declining. Most frequent in N and at moderate elevations; more local and montane in S, absent from very dry regions. Common migrant in Peninsula, Balearics and Canary Is.

♂
♀
Subalpine Warbler
Lesser Whitethroat
Spectacled Warbler
♀
♂
♀
Greater Whitethroat
♂

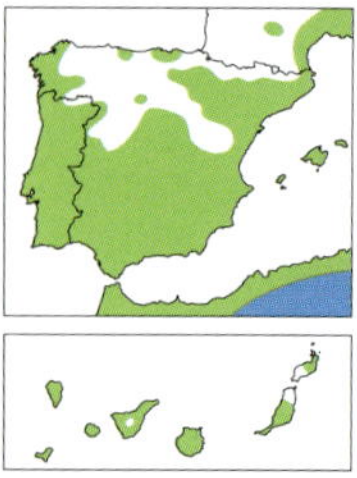

Sardinian Warbler *Sylvia melanocephala* (curruca cabecinegra). ~13 cm. Cap includes eye (as in Orphean Warbler); black in male, grey in female and juvenile male, browner in juvenile female. Resembles other pale-legged, tail-cocking, small scrub warblers but lacks rufous wing panel. Call: a loud rattling: *tcha–tcha–tcha*. Throat white, eye-ring red, tail graduated. Juvenile resembles those of other scrub warblers but darker, throat dusky, iris dark and orbital ring pale. Often sings in flight as well as perched; song similar to Subalpine Warbler, a series of rapid musical warblings interspersed with harsh notes ✦ Scrub (lentisc, scrub-oaks, broom, gorse) and tamarisk clumps, hedges, gardens, etc. ✦ Resident and dispersive. Sensitive to cold snaps ✦ Very common (a few million); most abundant in E, S and Balearics; rare towards N and at high elevations. Ssp. *leucogastra*: common and well distributed in Canary Is; **DD**

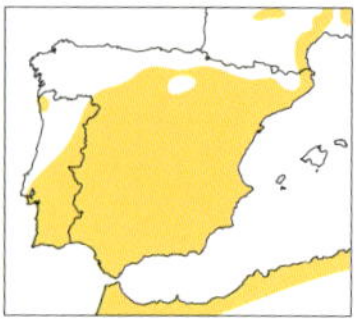

Orphean Warbler *Sylvia hortensis* (curruca mirlona). ~15 cm. Large, hooded; adults have conspicuous pale iris. Hood black in male, grey in female, brownish in juvenile. Throat white, whitish in juvenile; bill robust, legs dark; outer rectrices white. Adult male pinkish tinted below. Song loud, musical, with thrush-like notes repeated 3–4 times in short bursts: *chewree–chewree–chewree*. Calls: *chec*, *trr*… Forages in tree canopy, secretive ✦ Mediterranean woodlands, often open and low (Holm, Cork or Pyrenean Oak, Juniper, Aleppo Pine) or dehesas; also olive groves, hedges, orchards ✦ Trans-Saharan migrant (late Mar/May to Aug/Oct) ✦ Uncommon (a few hundred thousand) latterly increasing; absent from Biscay coastlands and deforested areas. Passage inconspicuous. Occasional or very rare in Balearics. Vagrant to Canary Is.

TREECREEPERS (Passeriformes, Certhiidae) – Small woodland birds. Creep up trunks, propped up on their tails. Eat insects and their larvae and pupas, and spiders. Eurasia and N America.

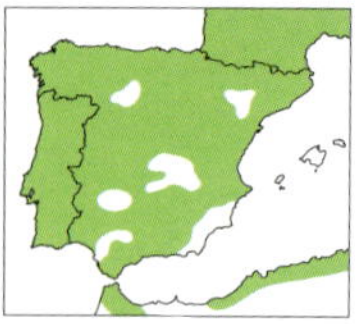

Short-toed Treecreeper *Certhia brachydactyla* (agateador europeo). ~13 cm. Common. Easily identified where Eurasian Treecreeper is absent. Small, mouse-like, with long, curved bill; climbs tree trunks. Song distinctive, brief (1–1.5 s) and emphatic: *ti–ti–titeroy–tí*; usual calls drier and louder than those of Eurasian: *siit*, *tsui*, *chink* ✦ All woodland and plantation types, riparian woods, dehesas, olive groves, parks. Nests in crevices and holes in trees, or in nestboxes ✦ Sedentary ✦ Ssps. *megarhynchos* (NW) and *brachydactyla* (elsewhere): very common (a few million) and widespread; increasing; most common in rainy regions and in closed mature forests.

Eurasian Treecreeper *Certhia familiaris* (agateador euroasiático). ~13 cm. In Pyrenees, Cantabrian Range and N Iberian Range. Very similar to Short-toed Treecreeper but whiter underparts contrast more with upperparts; supercilium more evident ahead of eye. Song allows rapid identification (although the two species imitate each other where they come in contact); song longer (≥3 s), with introductory notes (*sui*, *sui*, *sui*), a tremulous trill recalling a Goldcrest, and a more distinct, lower pitched ending recalling a Blue Tit. Usual call: a sharp thin *srii* or *tsii* ✦ Humid mature woodlands (beech, spruce, birch, etc.), mainly above 1,000 m ✦ Sedentary ✦ Ssp. *macrodactyla*: uncommon (a few tens of thousands of pairs) and restricted to montane forests.

Sardinian Warbler
♂
♀
Orphean
Warbler
♀
♂
Short-toed
Treecreeper
Eurasian
Treecreeper

NUTHATCHES (Passeriformes, Sittidae) – Insectivorous forest or rupestral birds that forage on tree trunks or rocks. Also eat seeds and nuts in winter. Eurasia and N America.

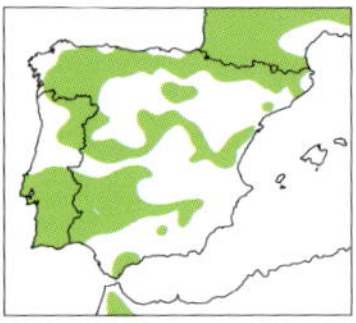

Eurasian Nuthatch *Sitta europaea* (trepador azul). ~14 cm. Broad black eyestripe descends neck sides. Blue-grey above, flanks rusty; bill large, tail very short (not used for support); toes and claws strong. Climbs head-first both upwards and downwards. Calls loud: *chuit–chuit* and many others. Song a series of repeated notes. Juvenile duller ✦ Woods and areas with trees, chiefly mature broad-leaved. Nests in holes, often woodpecker holes, or nestboxes ✦ Sedentary, with short winter movements ✦ Ssps. *caesia* (Pyrenees and Cantabrian range) and *hispaniensis* (elsewhere): common (hundreds of thousands of pairs), much more frequent in N and mountains, absent from deforested regions.

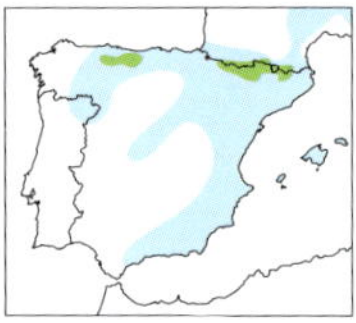

Wallcreeper *Tichodroma muraria* (treparriscos). ~16 cm. Mountains. Unmistakable. Small, grey, creeps nuthatch-fashion over and under rocks and up cliff faces, often revealing impressive carmine-red of its rounded wings. Outer primaries have large white spots. Flight jerky, recalls a small Hoopoe. Bill long, slender and decurved; tail short. Breeding male has black throat. Song is a series of loud whistles ✦ Nests in rock faces at high altitudes, usually on limestone. More widespread in winter. Insectivorous ✦ During Oct–Mar occurs at lower altitudes ✦ Scarce (hundreds or a few thousand) and restricted to highest sectors of the Pyrenees (especially in Huesca) and Cantabrian Range (Picos de Europa). More widespread in winter but generally very scarce, most frequent then in limestone mountains of N and E. Vagrant to Balearic Is.

WRENS (Passeriformes, Troglodytidae) – American family of small insectivores with thin, curved beaks. Allied to treecreepers and nuthatches. Only one species in Eurasia.

Northern Wren *Troglodytes troglodytes* (chochín común). ~10 cm. Tiny, barred, brown, with very small, often cocked tail. Movements mouse-like. Song surprisingly loud: a rapid sequence of trills and warblings, always identical in structure and duration (~5 s). Call: a sharp, persistent *tec, tec, tec*… Long white supercilium. White spots on median coverts ✦ Woodlands, lush scrub, hedges, shady rocky slopes, riparian vegetation ✦ Spanish birds sedentary or have altitudinal movements; very sparse wintering by European birds ✦ Ssps. *troglodytes* (Peninsula) and *kabylorum* (Balearics and perhaps S Peninsula): breeds very commonly (a few million), abundant in N, mostly in mountains in S; more widespread in winter.

DIPPERS (Passeriformes, Cinclidae) – Birds of mountain streams in the Americas (three species) and Palearctic (two) that capture insects underwater.

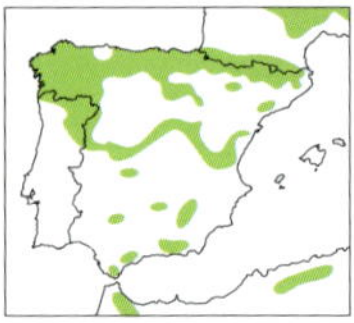

White-throated Dipper *Cinclus cinclus* (mirlo acuático europeo). ~18 cm. Blackish with large, striking, white bib; plump and short-tailed. Flies low over rivers, giving metallic calls. Perches on shoreline or rocks, where often bobs on legs and may display white eyelids. Lower breast reddish brown (ssp. *aquaticus*) or blackish-brown (*cinclus*). Juvenile grey, barred. Often solitary. Song: a rippling trill ✦ Clean, fast-flowing streams and rivers, generally in mountains, at up to >2.500 m. Nests between rocks, often in walls of bridges or watermill or behind waterfalls, etc. ✦ Sedentary ✦ Ssps. *cinclus* (N and W) and *aquaticus* (E and S): scarce (a few thousand pairs), relatively common in N and parts of Central Range, very local in E and S.

Wallcreeper

Eurasian Nuthatch

Northern Wren

White-throated Dipper

STARLINGS (Passeriformes, Sturnidae) – Medium-sized, noisy, gregarious passerines with strong bills and feet. Diets varied. Old World, mainly Africa and S Asia.

Spotless Starling *Sturnus unicolor* (estornino negro). ~21 cm. Short tail; triangular wings. Walks very upright. Bill slender and pointed, yellow with blue base in breeding male and pinkish in female, but blackish in juvenile. Breeding birds, especially males, glossy with purple sheen, unspotted, with shaggy throat and breast feathers. Small pale spots below in winter, soon lost through wear. Juvenile brown, darker than juvenile Common Starling, throat pale. Highly vocal year-round, with loud whistles and mimicry. Highly gregarious. Often gathers on roofs and wires ✦ Forages in farmland and pastures, nesting under roof tiles and in tree holes ✦ Resident ✦ Abundant (millions of pairs) and increasing); very widespread. Nests in Ceuta, Melilla and Chafarinas Is. Vagrant to Balearic Is.

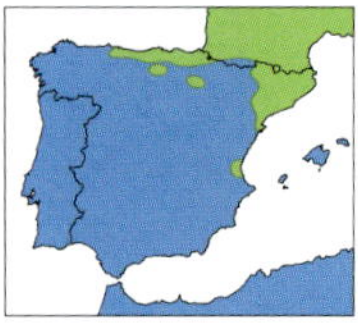

Common Starling *Sturnus vulgaris* (estornino pinto). ~21 cm. Unlike Spotless Starling retains some spotting in green-glossed breeding plumage; legs brownish-red. Looks paler, brownish in winter; when heavily spotted on head and body. Juvenile pale grey-brown, with pale throat, soon moulting into winter plumage, on body at first. Song similar to Spotless, with frequent gurgling, whistles less loudly. Often in enormous flocks ✦ Habitat as Spotless Starling. In winter frequents olive groves and orchards, roosting in reedbeds and dense clumps of trees, often in towns ✦ Spanish birds resident; wintering of European birds (Oct/Nov–Feb/Mar) ✦ Common (hundreds of thousands), and spreading; first colonised the Peninsula in the 1960s. Abundant in winter (millions), particularly in coastal regions and the Balearics. Present in Tenerife and Gran Canaria.

THRUSHES (Passeriformes, Turdidae) – Allied to Chats but somewhat larger, with robust legs. Quite frugivorous, especially in winter. Many are migratory. Worldwide.

Common Blackbird *Turdus merula* (mirlo común). ~25 cm. Familiar. Adult male black, bill and eye-ring yellow; 1st winter male has bill black and wings brownish-black. Female brownish-black, throat whitish, bill dark or yellowish-brown. Juvenile more rufous on body, pale-spotted above. Sings from winter; song a measured series of musical flutings, tailing away into twitters. Calls varied: *chook*, *chink–chink–chink*, *sriii…*, including a ringing alarm. Hops and runs ✦ Woods and scrublands, riparian growth, orchards, parks and gardens ✦ Resident, with wintering of European birds (mainly Oct–Feb) ✦ Ssp. *merula*: abundant (millions of pairs) and widespread, except in deforested regions. Ssp. *cabrerae*: C and W Canary Is, widespread and abundant; **DD**

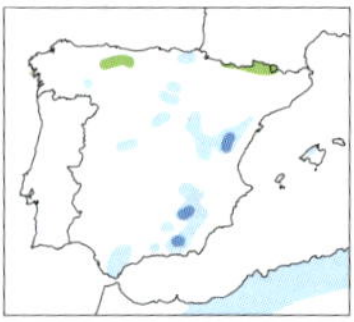

Ring Ouzel *Turdus torquatus* (mirlo capiblanco). ~25 cm. Mountains. White half-moon on breast diagnostic, although brownish-white in adult female and absent in juvenile and 1st winter female. Wing feathers edged pale above. Race *torquatus* largely black below. Scaly appearance more general in winter plumage. Juvenile has pale neck patch and pale barring below. Song a series of repeated short phrases. Calls: *chook–chook–chook*; a metallic *tac–tac–tac*; *chuirr* in flight. Forms winter flocks ✦ Nests mainly in open, subalpine, Black Pine woodlands. Winters in Juniper woods ✦ Migrant, with passage (Mar–Apr and Sept–Nov) and some wintering of birds from Britian and Scandinavia (ssp. *torquatus*) and the Alps (*alpestris*) ✦ Ssp. *alpestris*: scarce breeder (a few thousand) in Pyrenees and locally in Cantabrian and North Iberian ranges. Passage most frequent in E and Balearics. Winters mostly in S Aragón and Andalucía. Occasional in Canary Is.

Spotless Starling
summer
winter
summer
winter
Common Starling
Common Blackbird
♀ ad.
♂ ad.
Ring Ouzel
♀ ad.
♂ ad.

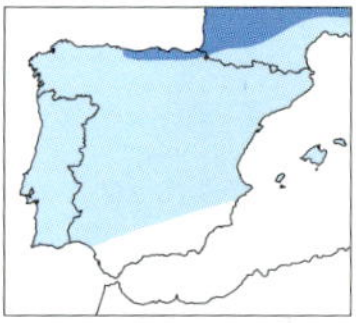

Fieldfare *Turdus pilaris* (zorzal real). ~25 cm. Shares white underwings, axillaries and underwing coverts with Mistle Thrush, but somewhat smaller and much darker. Colour contrast unique: head, rump and lower back grey; back reddish brown, streaked breast yellowish-rufous, tail black. Throat streaked. Flanks have black chevron markings. Noisy, harsh call: *chack–chack–chack*, often given in flight; also a thin *seeee* in flight. In flocks, often with other thrushes ✦ Selects open countryside with trees ✦ In winter (Nov–Feb). Irregular (with irruptions). Originate from Scandinavia, Finland and also C Europe ✦ Uncommon or scarce, most frequent in Biscay coastlands and Iberian range. Scarce in Balearics. Vagrant to Canary Is.

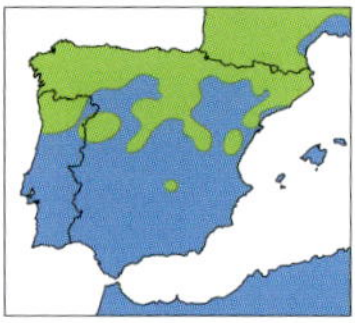

Song Thrush *Turdus philomelos* (zorzal común). ~22 cm. Most familiar spotted thrush. Underwings buff. Darker and warmer-toned than similar Mistle Thrush; also smaller and more compact. Flight call distinctive: a brief *tsip* or *chip*, often heard from migrating flocks at night. Juvenile speckled yellow on mantle and scapulars. Song, often given from a high tree perch, resembles Blackbird but characteristically repeats phrases 3–4 times. Discreet, forages on ground near cover. Smashes snail shells on stone 'anvils' ✦ Woods, riparian vegetation, humid open country with trees. Often frugivorous in winter, wintering in olive groves and Mediterranean scrubland ✦ Partial migrant, with passage and wintering (Oct/Nov–Feb/Mar) of European birds, including some from the British Isles and vicinity (ssp. *clarkei*) ✦ Fairly common breeder (hundreds of thousands), apparently spreading. Frequent in the humid N but scarce in S and C, where confined to mountains. Abundant in winter (several million), especially in the coastal N and E, Andalucía, Extremadura and Balearics; much hunted. Scarce but regular in Canary Is.

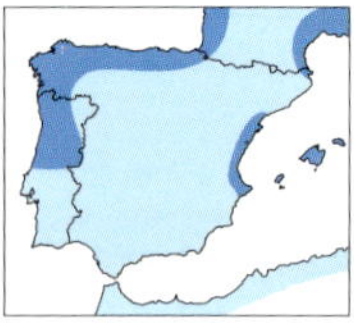

Redwing *Turdus iliacus* (zorzal alirrojo). ~21 cm. Our smallest thrush; with bold white supercilum and red underwings. Characteristic flight call: a sharp *seeep* often heard from nocturnal migrants in November. Reddish flank, heavily streaked flanks and breast and very white belly apparent when perched. 1st winter has white tips to tertials and outer greater coverts. In flocks ✦ Olive groves, Mediterranean woodlands, forest rides and clearings. Very frugivorous in winter ✦ Wintering and passage (late Oct/Nov–Feb/Mar) of birds from Scandinavia, Finland and Baltic States (ssp. *iliacus*) and Iceland (ssp. *coburni*) ✦ Common (many hundreds of thousands, variable), but most frequent in coastal N and E, Andalucía and Balearics. Scarce and irregular in Canary Is.

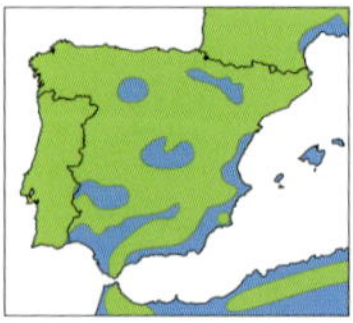

Mistle Thrush *Turdus viscivorus* (zorzal charlo). ~25 cm. Large, underwings white. Typical call: a loud grating *charrrrrr*. Pale and rather uniform grey-brown above; underparts with clear, round black spots; face looks pale, with dark and pale cheek markings; tertials and upperwing coverts edged pale; outer tail feathers tipped white. Juvenile distinctively pale spotted above. Song, heard from early in year, recalls Blackbird but simpler, with short, loud, ringing phrases. Often in groups, especially post-breeding ✦ All types of open woodlands (pinewoods, junipers, beechwoods…) or dehesas; also in riparian trees and open countryside with trees. Frugivorous in winter (mistletoe, juniper…) ✦ Partial migrant. Spanish birds resident or wanderers, with poorly evident passage and wintering of C European birds (mainly Nov–Feb) ✦ Quite common (hundreds of thousands of pairs), most frequent in regions of some elevation, particularly in the S. Fairly common on passage and in winter in Balearics. Vagrant to Canary Is.

Fieldfare
Song Thrush
Redwing
Mistle Thrush

CHATS AND FLYCATCHERS (Passeriformes, Muscicapidae) – Small insectivores formerly included in the thrush family. Fine-billed and large-eyed, with erect posture. Mainly Old World

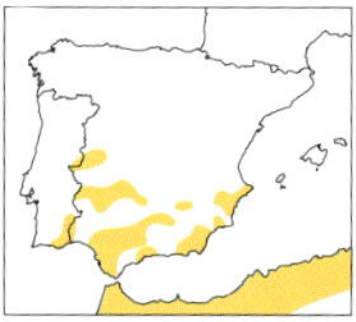

Rufous Scrub-robin *Cercotrichas galactotes* (alzacola rojizo). ~16 cm. In S. Long, rufous tail, tipped black and white, is often fanned, cocked or laid over the back. Superciliar and eye stripes noticeable. Legs pale. Juvenile similar, fine-spotted. Call: a harsh *tek, tek*. Sings from perch or in parachuting display flight, tail-spread; phrases musical and varied but brief and well-spaced ✦ Woody crops (olives, vines, almonds, citrus fruits) and riparian scrub in dry regions ✦ Summer visitor (Apr/May–Aug/Sept), trans-Saharan migrant ✦ Uncommon (tens of thousands or a few hundred thousand), strongly declining. Vagrant to Balearic and Canary Is. **EN**

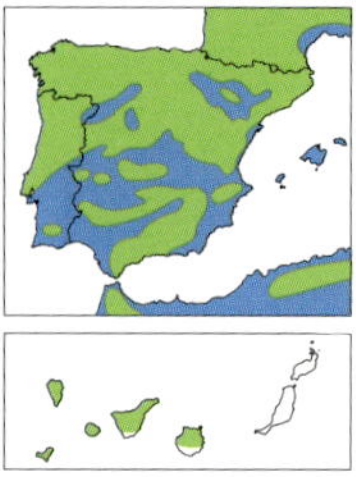

European Robin *Erithacus rubecula* (petirrojo europeo). ~14 cm. Face and breast orange-red. Juvenile brown, pale-spotted. Call: a distinctive *tic–tic–tic–tic*. Song, heard most of year, a sweet succession of melodious trills, often high-pitched. Chunky, bill fine, head and dark eyes large, breast rounded, belly whitish. Hops around on ground when foraging and uses low perches. Solitary ✦ Woodland, riparian growth, open woodland with trees; frequently in lusher Mediterranean hill woodlands and in parks and gardens ✦ Partial migrant, with wintering and passage of birds from C and N Europe (~Sept/Oct–Feb/Mar) ✦ Ssp. *rubecula*: common (a few million), very common in humid coastlands and mountains of the N half, largely in mountains in the S. Abundant and widespread on passage and in winter throughout the Peninsula and Balearics; scarce in W Canary Is. Ssp. *melophilus*: scarce winterer (British Isles). Ssps. *superbus* (Tenerife, **DD**), *marionae* (Gran Canaria) and *microrhynchus* (La Palma, La Gomera and El Hierro): common, sedentary.

Common Nightingale *Luscinia megarhynchos* (ruiseñor común). ~16 cm. Very secretive but revealed by its powerful, varied and musical song, heard day or night. Song includes repeated elements: a slow *peoo–peoo–peoo…*, a rapid *chook–chook–chook*, and others. Alarm call: a frog-like *krrrrr*. Reddish tail and rump are most noticeable plumage features. Juvenile spotted. Well proportioned, alert-looking and upright in posture ✦ Dense humid undergrowth, tall dense scrub, riparian thickets, orchards, shady gardens ✦ Summer visitor, trans-Saharan migrant (~late Mar/May–Aug/Sept) ✦ Very common (many hundreds of thousands or a few million); most frequent in Mediterranean regions that are not too dry; strongly associated with watercourses in S half; absent from Biscay coastlands and high mountains. Uncommon but regular on passage in Canary Is.

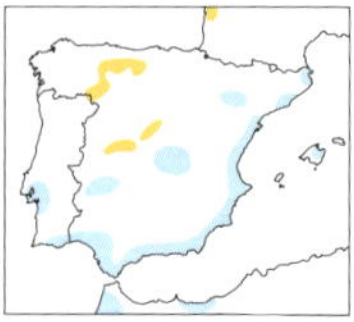

Bluethroat *Luscinia svecica* (ruiseñor pechiazul). ~14 cm. Showy blue throat and upper breast of male, sometimes with a white central spot (red in vagrant ssp. *svecica*). Female has white throat, bordered black, juvenile spotted. All have distinctive tail pattern, with bright chestnut tail sides, and white supercilium. Tail often cocked. Song varied and mimetic, with repeated elements: *ting–ting–ting… terr–terr–terr…* ✦ Montane scrub (junipers, heathland) at about 1,000–2,000 m. Winterers favour marshland vegetation ✦ Migrant, most trans-Saharan, with many winterers; passage of European birds (~Sept/Oct and Mar/May) ✦ Ssp. *azuricollis*: uncommon (thousands of pairs) restricted to Central and Cantabrian ranges (has bred Iberian range). Ssps. *cyanecula* and *namnetum* (mainly from Low Countries and W France): evident on passage; scarce and local in winter, most frequent in coastal E and S. Vagrant to Canary Is.

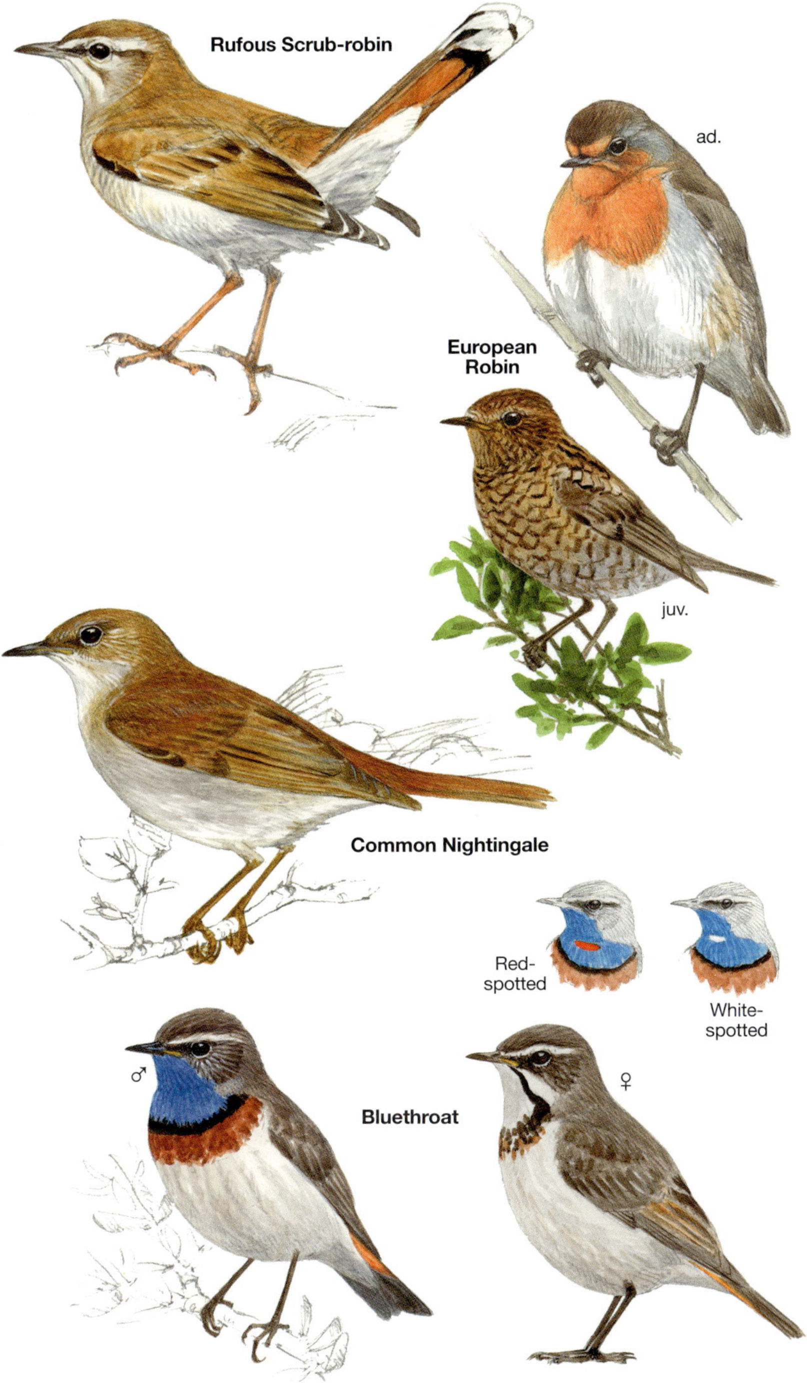
Rufous Scrub-robin
ad.
European
Robin
juv.
Common Nightingale
Red-
spotted
White-
spotted
♂
Bluethroat
♀

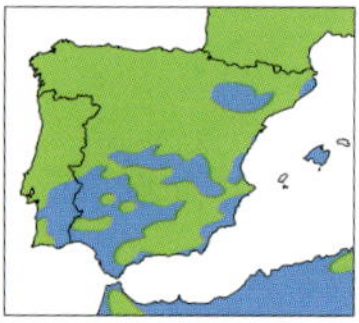

Black Redstart *Phoenicurus ochruros* (colirrojo tizón). ~14 cm. Males variably dark, darkest in worn spring plumage and in ssp. *aterrimus*; least dark in autumn, ssp. *gibraltarensis* and 1st years; wing panel white. Female darker and greyer than female Common Redstart. Juvenile resembles female, with some slight breast barring. Sings from a high perch, on a rock or roof; a rapid series of sharp notes ending in a rattle. Calls: *tsip*, *tic–tic–tic*. Often hunts from perches or hopping around on the ground; perches upright, often vibrating tail ✦ Rocky areas, from sea level to high mountaintops; also villages, large buildings, ruins. Winters in open country ✦ Spanish birds resident or perform seasonal movements; wintering and passage of European birds (~Oct/Nov and Feb/Mar) ✦ Ssps. *gibraltarensis* (N and European migrants) and *aterrimus* (C and S): common breeder (hundreds of thousands of pairs), widespread and very frequent in N but only in mountains in S. Very common in winter, when many descend to the coasts; similarly in Balearics. Scarce in Canary Is.

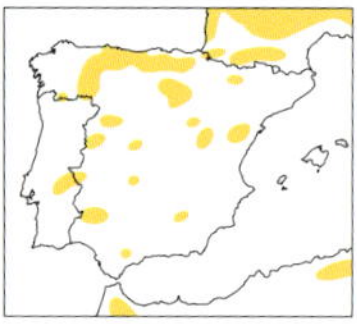

Common Redstart *Phoenicurus phoenicurus* (colirrojo real). ~14 cm. Male showy. Female much paler, browner and warmer-toned than female Black Redstart, throat white, breast tinged orange. Juvenile spotted like juvenile Robin but with red tail. 1st year male lacks white on forehead and is black only on throat. Song: a melancholic but pleasing repetition of brief elements, starting with sharp, separate notes. Calls: *huit*, *huit–tic–tic*. Behaviour as Black Redstart ✦ Areas with mature trees offering nest holes: open oakwoods, dehesas, pinewoods with nestboxes, large gardens ✦ Summer visitor, trans-Saharan migrant with passage of birds from C and N Europe (~late Mar–May and late Aug–Oct) ✦ Scarce (tens of thousands of pairs), declining, chiefly in wooded regions at some altitude; **VU**. Common on passage, including in the Balearic and Canary Is.

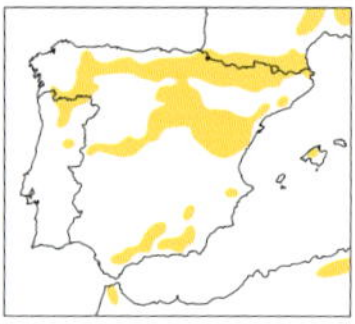

Common Rock-thrush *Monticola saxatilis* (roquero rojo). ~19 cm. Mountains, in spring/summer. Intermediate between a *Turdus* thrush and a redstart in size, appearance and behaviour. Breeding male very showy. Other plumages brown and vermiculated, always with red tail. Autumn male very barred, retaining a suggestion of the spring pattern. Juvenile drabber than adult female; back paler and spotted. Song fluting and warbling, often given in display flight. Calls: *tak*, *chac–chac* ✦ Rocky areas with nearby pastures or open scrub, frequently at 1,000–2,000 m. Nests under rocks or in cavities in walls, byres, ruins ✦ Summer visitor (~Apr–Sept/Oct), winters in tropical Africa ✦ Scarce (thousands of pairs), commoner in N half. Nests locally in Mallorca. Vagrant to Canary Is.

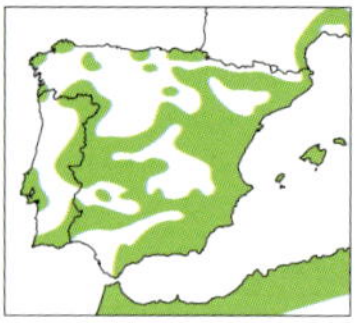

Blue Rock-thrush *Monticola solitarius* (roquero solitario). ~22 cm. Rocks. At a distance or in a brief view recalls a small, short-tailed Blackbird with a long bill tilted upward. Intense pale blue of male only apparent in good light. Female darker than female Common Rock-thrush and lacks red tail. Juvenile like female, with some spotting above. Song, given from perch or in flight, a sequence of loud, musical, whistling phrases (not to be confused with that of Black Wheatear, often found in same areas). Characteristically spends long solitary periods on conspicuous rocky perches ✦ Cliffs and rocky scree slopes from sea level to mid elevations; sometimes on castles, churches, ruins, bridges. More widespread in winter ✦ Resident, with local seasonal movements and dispersion ✦ Uncommon (tens of thousands); most frequent in E and S, rare and local in N. Nests in Ceuta, Melilla and Chafarinas Is. Vagrant to Canary Is.

Black Redstart
♀ ad.
♂ ad. summer
Common Redstart
♀ ad.
♂ ad. summer
Common Rock-thrush
♂ ad. summer
♀ ad.
Blue Rock-thrush
♂ ad.
♀ ad.

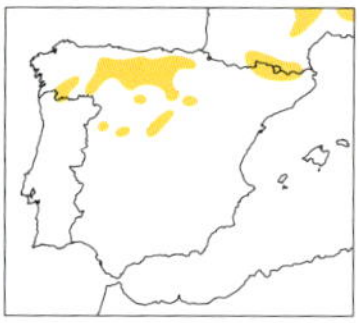

Whinchat *Saxicola rubetra* (tarabilla norteña). ~13 cm. Differs from Common Stonechat in having broad white supercilia and white at tail base. Cap streaked; face and ear coverts blackish in male, brown in female; throat, breast and flanks variably orange-buff; wings show white on inner coverts, especially in male, which also has white on primary coverts. Juvenile spotted above. Call: *tec–tec*, *tui–tec–tec*, gentler than Stonechat; song slower, varied and musical, with mimicry. Hunts from perches on shrubs, posts and wires ✦ Meadows and pastures with scattered shrubs or small trees, sometimes in croplands, frequently at 1,000–2,000 m ✦ Summer visitor, with passage (late Mar–May and late Aug–Oct) of birds from C and N Europe ✦ Scarce (tens of thousands); chiefly in N Castilla y León. Common on passage, but strongly declining. Common on passage in Balearics; scarce migrant in Canary Is.

Common Stonechat *Saxicola torquatus* (tarabilla común). ~12 cm. Male unmistakable, although duller in winter. Female brown-headed, duller below with only a suggestion of white half-collar. Juvenile pale-throated, spotted on head and back. Tail all dark; rump paler or whitish. Typical call suggests two pebbles struck together: *chac–chac*, *wit–chac–chac*. Song a brief, clear, jingling warble. Hunts from open perches (on boulders, shrubs wires); perches very upright, flicking wings and tail. Often in pairs or family parties ✦ Open scrubland and pastures with shrubs, at variable gradients and altitudes (rare >1,500 m). Also field margins and young plantations ✦ Spanish birds partly migratory; wintering and passage (~Mar/Apr and Sept/Oct) of European birds ✦ Ssps. *hibernans* (NW) and *rubicola* (elsewhere): common (hundreds of thousands or some millions) and widespread, except in very arid or intensively cultivated regions. Common in winter, especially in SW and coastal regions. Scarce and irregular in Canary Is. Ssps. *maurus/stejnegeri* (Asia): vagrants.

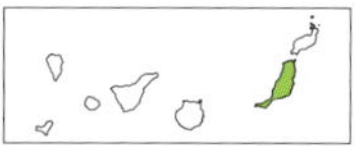

Fuerteventura Stonechat *Saxicola dacotiae* (tarabilla canaria). ~12 cm. Endemic to Fuerteventura, Canary Is. White throat and supercilia recall Whinchat. Male has orange only on breast, intensity variable. Female and juvenile very dull, latter spotted. Tail entirely dark, upper tail coverts greyish; only white on inner wing coverts of male apparent in flight. Song, often in flight, recalls Common Stonechat. Call: *tec, tec* ✦ Hillsides, gullies and badlands with some shrub cover; also on edges of fields and in gardens ✦ Sedentary ✦ Ssp. *dacotiae*: fairly common (hundreds of pairs) and widespread, but declining; **EN**. (Ssp. *murielae*: extinct, formerly on Alegranza and Montaña Clara islets).

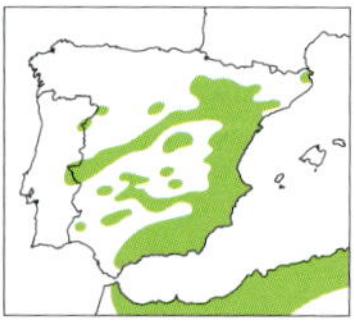

Black Wheatear *Oenanthe leucura* (collalba negra). ~17 cm. Large and unmistakable. Male glossier; female browner, juvenile has brown wings Call: *pii–pii–pii*, *chac*. Melodious song, often given in flight, more continuous than in other wheatears, suggesting Blue Rock-thrush (but less varied and harsher). Displays with fanned wings and tail. Moves around between rocks and boulders, concealing itself with ease ✦ Rocky or open sparsely-vegetated escarpments in dry regions. Nests in holes, caves, abandoned farm buildings ✦ Mainly sedentary, although moves away from some inland and mountain areas in winter ✦ Ssp. *leucura*: scarce (thousands of pairs) and quite local, most frequent towards E and S and especially in the arid SE (Almería, Murcia); declining, especially in N. Ssp. *syenitica*: breeds Melilla. Vagrant to Balearic Is.

♀ ad.
Whinchat
♂ ad. summer
♀ ad.
♂ ad. summer
Common Stonechat
Fuerteventura Stonechat
♂ ad. summer
♀ ad.
♂ ad.
Black Wheatear
♀ ad.

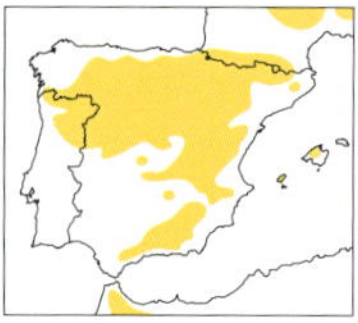

Northern Wheatear *Oenanthe oenanthe* (collalba gris). ~15 cm. Black terminal tail band and central rectrices contrast with white tail sides and rump, as in other wheatears. Breeding male has cap, nape and back grey. Other plumages browner and resemble those of Black-eared Wheatear although paler. Autumn males retain white supercilium and blackish mask. Juvenile buff spotted and with dark feather edging on breast and back. Call stonechat-like: *chac*, *wit–chac*. Song a brief warble, with lark-type and other chirpy notes. Hunts from perches or running on ground ✦ Treeless areas of short pasture or low open scrub, with rocks, walls or rabbit burrows for nesting. In mountains, fallow areas and cereal croplands. Often on tilled land and coastal grasslands on passage ✦ In summer with passage (~Mar–May and late Aug–Oct) of birds from C and N Europe (ssp. *oenanthe*) and Iceland and Greenland (*leucorhoa*, larger and more colourful) ✦ Ssps. *oenanthe* (Pyrenees) and *libanotica* (elsewhere): fairly common breeders (hundreds of thousands of pairs); well distributed in N half but only on mountains in S; scarce breeder in Ibiza and sites in Mallorca. Very common on passage in Peninsula and Balearics and fairly common migrant in Canary Is.

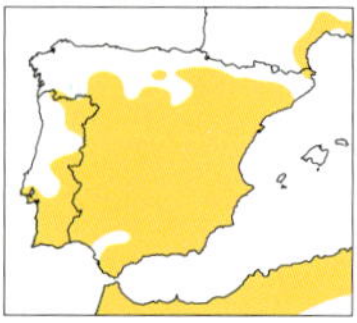

Black-eared Wheatear *Oenanthe hispanica* (collalba rubia). ~14 cm. Two morphs: white-throated and black-throated, little apparent in females. Tail more extensively white than in Northern Wheatear but sides partly black. Male variably coloured above, from ochre to very pale cream. Female darker-winged, more tawny above and with darker ear coverts than female Northern. Sings from perch or in flight with well-spaced, brief (1–1.5 s) chirping phrases, recalling Greater Short-toed Lark. Hunts from perches on shrubs or stones ✦ Open Mediterranean scrub, also clearings and open areas on hillsides, woody crops, fallows. Nests under shrubs ✦ Trans-Saharan migrant (late Mar/Apr–late Aug/Oct) ✦ Fairly common (hundreds of thousands of pairs), declining; restricted to Mediterranean climatic region, most frequent towards S and E; scarce in Balearics; **NT**. Scarce and irregular on passage in Canary Is. Ssp. *melanoleuca* (SE Europe): vagrant.

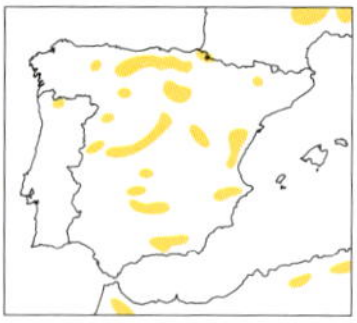

European Pied Flycatcher *Ficedula hypoleuca* (papamoscas cerrojillo). ~13 cm. Unstreaked, bicoloured, with white wing patches (tertial edges and greater coverts). Smaller than Spotted Flycatcher. Often flicks one wing upwards after landing. Often calls: a metallic *huit* or *huit–tic*. Breeding male essentially black and white, with large white wing patches, white forehead and a hint of a white collar (often almost complete in Spanish race). Male **Collared Flycatcher** *Ficedula albicollis* (papamoscas acollarado), rare migrant (p. 237), has a broad, distinct white collar and a whitish rump. Female and autumn male brown and white, tail blacker in male. Juvenile spotted above with dark feather edging on throat and breast. 1st autumn birds show pale edges to some median coverts. Song: a short sequence of notes, alternating in pitch and often ending in an abrupt trill *si-tsu-tsi-tsu-tsi-tsu... tsrr-tsirr-tsirr* ✦ Montane woodlands (oaks, beeches, pines) with nestboxes or tree holes for nesting, generally at 1,000–1,500 m. On passage anywhere with trees or shrubs, very frequent in parks and gardens ✦ Trans-saharan migrant, with passage (Apr–May and Aug–Oct) of birds from C and N Europe ✦ Ssp. *iberiae*: fairly common breeder (a few hundred thousand) and perhaps increasing, but local; commoner towards the N. Ssp. *hypoleuca*: abundant on passage in the Peninsula, more so towards the W in autumn and the E in spring, and in the Balearics. Common migrant in Canary Is.

Northern Wheatear
♂ ad. summer
♀ ad.
Black-eared Wheatear
♂ ad. summer
♀ ad.
♀
♂
European Pied Flycatcher

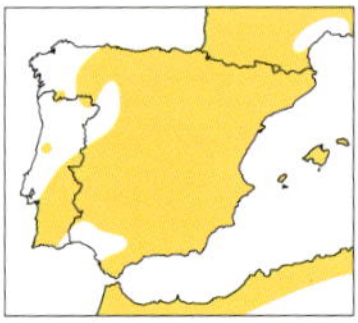

Spotted Flycatcher *Muscicapa striata* (papamoscas gris). ~14 cm. Greyish brown, with streaked breast, throat and cap. Characteristically returns to the same perch after flying out to catch a flying insect, often with an audible snap. Call: *sit* or *tsi*. Slender; forehead sloping, crown feathers sometimes raised; bill longish and broad; legs very short, tail long, wings pointed. Juvenile unstreaked but pale-spotted above. In autumn, adults are in worn plumage but juveniles are fresh-plumaged, with pale edged tertials and upperwing coverts. Song audible only at close range, a high pitched series of well-spaced squeaks and whistles ✦ Diverse wooded areas (woodland rides and clearings, orchards, riparian woods, gardens). Nests in sheltered sites ✦ Trans-Saharan migrant, with evident passage (Apr–May and Aug–Oct) of European birds ✦ Ssp. *striata*: breeds commonly (hundreds of thousands) although irregularly distributed, frequent in coastal regions but scarce and localised towards the interior and NW; nests in Ceuta and Melilla. Common on passage; when regular in Canary Is. Ssp. *balearica*: Balearics, abundant.

KINGLETS (Passeriformes, Regulidae) – Half a dozen species of woodland insectivores, traditionally included in family Sylviidae. Diminutive, with tiny, straight, slender bills. Eurasia and N America.

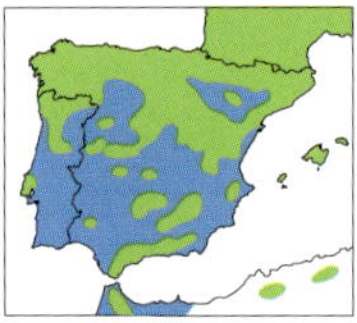

Common Firecrest *Regulus ignicapilla* (reyezuelo listado). ~9 cm. Common. Tiny, compact, greenish, with striped head. Calls very high-pitched: *si–si–si*. Differs from Goldcrest in having obvious white supercilium and black eye-stripe, brighter coloration and golden neckside patches. Song: a very high-pitched note similar to the call note repeated increasingly loudly and ending in a brief trill: *si–si–si–sisisisirrrr*. Erectile crest: orange in male, yellow in female, lacking in juvenile. Wings have two white wingbars and one black one. Moves restlessly through treetops, often hovering. Tends to accompany tit flocks ✦ Nests in conifer forests and plantations; also in evergreen and deciduous oakwoods, beechwoods, riparian woods. Widespread in winter in woods, dehesas, gardens, tall scrublands ✦ Partial migrant; with altitudinal movements of Spanish birds and wintering of C European birds. Passage (~Sept–Nov and Mar–Apr) ✦ Ssp. *ignicapillus*: common (a few million), increasing; frequent in N, more restricted to humid montane areas in C and S. Ssp. *balearicus*: very common and widespread in Balearics.

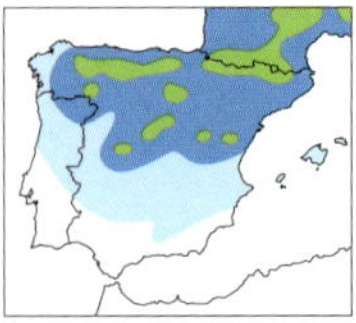

Goldcrest *Regulus regulus* (reyezuelo sencillo). ~9 cm. Nests in mountains. Lacks superciliar and ocular stripes; dark eye stands out in plain pale face. Black crown stripes do not meet on forehead, except in Canary Is races; median crown stripe orange in male, yellow in female, absent in juvenile. Calls very similar to Firecrest but even higher pitched. Song more complex, a brief repeated phrase ending in a final flourish: *pititlí–pititlí–pititlí…si-sisuriao* (see Eurasian Treecreeper, p. 184); song briefer in Canary Is, lacking final flourish. Behaviour as Firecrest, with which it often associates in winter ✦ Forests, nesting in pines, birchwoods, beechwoods, often at subalpine elevations, generally above 1,000 m. In winter also in Aleppo pinewoods, junipers, plantations, riparian woods, parks… In Canary Is occupies montane woodlands including mixed pinewoods, laurel forest and tall heathland, especially where the tree heath, in which it usually nests, abounds ✦ Partial migrant, with altitudinal movements of Spanish birds and wintering of C European birds (~Nov–Mar) ✦ Ssp. *regulus*: Peninsula, fairly common breeder (a few hundred thousand), but of limited distribution; in C only in Scots pine forests; fairly common and more widespread in winter, more so in N half; uncommon in Balearics. In Canary Is. (**DD**), ssps. *teneriffae* in Tenerife and La Gomera and *ellenthalerae*: in La Palma and El Hierro; generally common (some thousands of pairs) and widespread.

Spotted Flycatcher
juv.
Common Firecrest
♂
♀
Goldcrest
♂
♀

ACCENTORS (Passeriformes, Prunellidae) – A dozen species, chiefly montane. Brownish-plumaged, sexes alike. Eat insects, seeds, fruits. Eurasia.

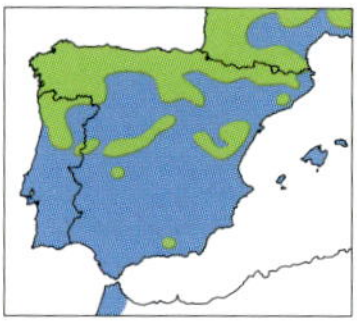

Dunnock *Prunella modularis* (acentor común). ~14 cm. Dull plumaged, head and breast grey, more so in male; brown, heavily streaked above and on flanks; juvenile yellower, also streaked on head and neck. Call distinctive: a sharp *siit* or *siip*. Song: an even-pitched succession of warblings, recalling Wren but less loud and briefer (~3 s) ✦ Scrub of lusher, non-arid, types; frequent at medium and higher elevations, abundant in broom scrub ✦ Partial migrant, with altitudinal movements and wintering of European birds (~Oct–Mar) ✦ Ssp. *mabotti*: fairly common breeder (hundreds of thousands), common at all altitudes in N but only in mountains in C and Sierra Nevada. Ssp. *modularis*: common winterer, mainly in E, S and Balearics.

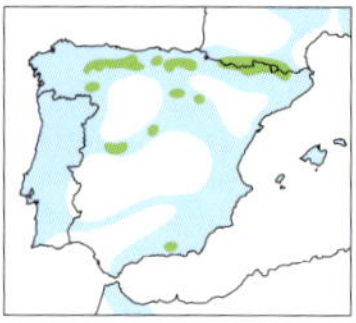

Alpine Accentor *Prunella collaris* (acentor alpino). ~17 cm. Mountains. Largish, greyish with rufous-streaked flanks. Bill base yellow; throat and chin white, barred black; upperwing coverts blackish, tipped white. Has loud lark-type calls: *chirr–rip*, *chirririrp*. Song: a varied, sustained warble, from perch or in flight. Often in small groups ✦ Nests generally above 1,800 m, in rocky areas, near pastures and snowfields; also around mountain refuges and ski stations. Winters mainly at lower altitudes, in rocky sierras down to sea level ✦ Resident, with altitudinal movements and wanderings (~Oct–Apr) ✦ Scarce (thousands of pairs) although common within its restricted habitat. Very local away from the Pyrenees and Cantabrian range. Scarce winterer in Balearics.

LAUGHING-THRUSHES AND ALLIES (Passeriformes, Leiothrichidae) – Small or medium-sized, with rounded wings, long tails and straight bills. Insectivorous and frugivorous. Africa and S Asia.

Red-billed Leiothrix *Leiothrix lutea* (leiótrix piquirrojo). 14–15 cm. Greyish, bill bright red, throat and breast yellow, wings with red and yellow areas. Tail forked. Female has cap greener, ear coverts greyer and wing panel smaller. Juvenile pale-billed. Alarm call: a guttural, slightly nasal, short, abrupt *shrik* or *shreep*. Often in small groups ✦ Woodlands with dense undergrowth, near streams ✦ Sedentary ✦ A common cagebird (the Pekin Robin); feral populations established near Barcelona since the 1990s; spreading (some hundreds).

WAXBILLS AND ALLIES (Passeriformes, Estrildidae) – Small, mainly tropical, seed-eaters, often strikingly coloured. Africa, S Asia and Australia.

Common Waxbill *Estrilda astrild* (estrilda común). ~11 cm. Small, grey-brown, with long graduated tail and short rounded wings. Adult (male especially) has bill and eye mask bright red, pale belly with central red patch, black undertail and fine barring on back and flanks. Juvenile duller, unbarred, bill brown. Call: *chip*; song trisyllabic: *dee–dee–JEE*. Often in flocks ✦ Marsh and riparian vegetation; also ricefields ✦ Sedentary but wanders ✦ African, introduced in Portugal in 1960s and now extends into Extremadura and W Galicia; also locally in parts of Andalucía, the Mediterranean coastlands, Mallorca and Canary Is.

Red Avadavat *Amandava amandava* (bengalí rojo). ~10 cm. Tiny. Breeding male dark red, spotted white below. Winter male and female brownish grey above, whitish below; uppertail coverts red, on male especially. Wings brown, tail and lores black. Juvenile lacks red uppertail coverts. Call: a somewhat metallic, high *teei* or *tsi*. Song: a high-pitched twitter, descending. Gregarious ✦ Irrigated farmland, in ditches, canals and damp areas with reedmace, reedbeds, etc.; also ricefields ✦ Sedentary ✦ S Asian, introduced birds established in the 1970s/1980s; spreading. Common in irrigated zones of Extremadura and part of Andalucía, with some presence elsewhere.

Dunnock
Alpine Accentor
Red-billed
Leiothrix
Common Waxbill
♂ ad.
♀ ad.
♂ ad. summer
Red Avadavat
♀ ad.

SPARROWS (Passeriformes, Passeridae) – Sombre-coloured seed-eaters with stout conical bills. Songs unremarkable. Gregarious. Sedentary. Old World.

House Sparrow *Passer domesticus* (gorrión común). ~15 cm. Ubiquitous although strongly tied to human settlements. Brown with streaked back. Male markings: black bib, grey crown, whitish cheeks, most apparent in worn spring plumage. Female and juvenile duller. Hops. Calls: *chirp*, *chip* and other similar notes, repeated as a slow series in the 'song' ✦ Cities, towns, villages, homesteads, farm buildings, etc. Nests under tiles or in holes in buildings or trees, in House Martin nests, within White Stork nests, etc., or constructs untidy, grassy, spherical nests in trees or shrubs, often nesting colonially, sometimes with Spanish Sparrows. Forages in croplands, streets, parks ✦ Sedentary, with dispersion ✦ Ssps. *domesticus* (Peninsula) and *balearoibericus* (Balearics) abundant (many millions of pairs) and well distributed. Has nested in Gran Canaria.

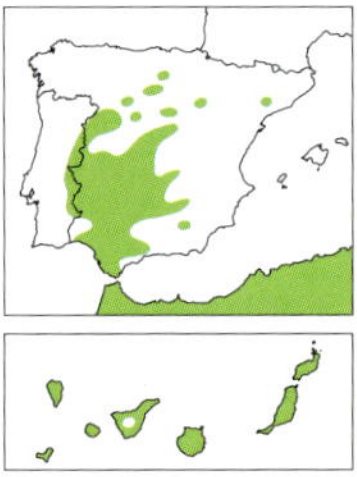

Spanish Sparrow *Passer hispaniolensis* (gorrión moruno). ~15 cm. Resembles House Sparrow, with which it may hybridise, but much less widespread. Male showy in spring with bright chestnut cap, very white supercilium and cheeks, black bib and breast and broad black streaks above and on flanks; markings duller in autumn and winter. Female as House Sparrow female, sometimes indistinguishable, generally with faint streaking on paler underparts ✦ Rural, nesting colonially, in trees and shrubs, colonies at times numbering hundreds or thousands of pairs, near crops or pastures, often in irrigated areas. Occupies towns and villages in Canary Is ✦ Resident, with dispersion and nomadism, sometimes long-distance (some reach N Africa) ✦ Common (hundreds of thousands of pairs), increasing; abundant in Extremadura and nearby areas, with scattered colonies elsewhere: range expanding. Vagrant to Balearic Is. Abundant in Canary Is.

Eurasian Tree Sparrow *Passer montanus* (gorrión molinero). ~13 cm. Smaller and more compact than other sparrows, sexes alike. Cap brownish-rufous, cheeks white with black ear covert spot; bib black, small; collar white. Juvenile duller, markings indistinct. Calls similar to House Sparrow, with a diagnostic *chek* or *cherk* given in flight. In small flocks, often with finches ✦ Nests on town outskirts, isolated buildings, in riparian woods and copses, parks, etc., using all types of cavities and often nestboxes ✦ Sedentary and dispersive, with some arrivals from other European countries ✦ Common (a few million) at variable densities: selects cultivated areas, often irrigated ones, and avoids mountains; declining. Scarce local breeder in Balearics and Gran Canaria.

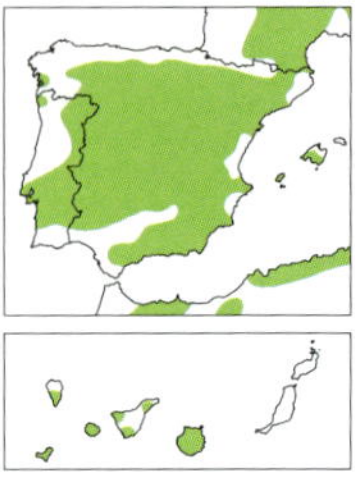

Common Rock-sparrow *Petronia petronia* (gorrión chillón). ~16 cm. Recalls a large, pale, female House Sparrow with striking pale crown and superciliary stripes. White spots on corners evident on spread tail. Calls *chewee* persistently. Adult has yellow spot at base of throat, hard to see. Sexes similar. Bill large, wings long, tail short. Walks rather than hops. Often in flocks ✦ Habitat variable but generally open and dry, often rocky. Small colonies inhabit escarpments, scree slopes, bridges, ruins, dehesas, etc. ✦ Sedentary, with dispersion and nomadism ✦ Common (hundreds of thousands of pairs) and widespread, more frequent in E half and absent or very local in Galicia and the Biscay coastlands. Common in Ibiza and Formentera, local in Mallorca. Uncommon and declining in Canary Is.

House Sparrow
♂ winter
♂ summer
♀
♀
Spanish Sparrow
♂ summer
winter
Eurasian Tree Sparrow
ad.
Common Rock-sparrow

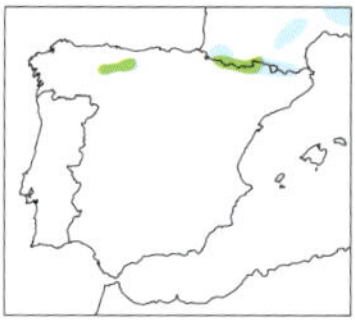

White-winged Snowfinch *Montifringilla nivalis* (gorrión alpino). ~18 cm. Pyrenees and Cantabrian range. Slender, with grey head, brown back and white underparts. Tail and wings show striking white markings in flight (compare Snow Bunting, p. 220). Bill black in summer, yellow in winter; small black bib most evident in male but indistinct in winter. Juvenile lacks black bib; head buffish, white wing patches buff-tinted. Calls: *chewik*; a penetrating *cheeáa* when taking flight; *prrrt* in alarm. Song sparrow-like, from perch or in flight, with repeated phrases: *si–ti–chi*… Highly gregarious after nesting ✦ High elevations, generally >1,800 m, often around screes, rocky outcrops and the edges of snowfields ✦ Sedentary, descending lower in winter; occasionally travels some distance; some reach the NE from the Alps ✦ Scarce (a few thousand), restricted to northern mountains. Recorded sporadically in winter in other mountain regions. Occasional in Balearics.

PIPITS AND WAGTAILS (Passeriformes, Motacillidae) – Slender terrestrial passerines, generally long-tailed, that often walk or run. Allied to finches. Bill fine, hindclaw long, tertials long, outer rectrices white. Insectivorous. Nest on the ground or between stones. Cosmopolitan.

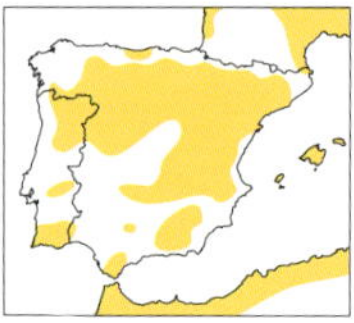

Tawny Pipit *Anthus campestris* (bisbita campestre). ~17 cm. Sandy, adult barely streaked, wagtail-like. Superciliar, loral, moustachial (short) and submalar (long and thin) stripes conspicuous. Median coverts form dark bar. Legs pinkish, long. Hindclaw curved, shorter than in Richard's Pipit. Juvenile show breast spotting and streaked mantle. Calls: a sparrow-like *cheep*; also *tseep*, recalling Yellow Wagtail. Song, given in undulating song-flight, a slowly repeated, metallic, *chur–lee* or *cheer–ree* ✦ Level areas with short grass and scattered shrubs, in *páramos*, dry montane areas, pastures, fallows, vineyards ✦ Summer visitor and migrant (~Apr/May–Sept/Oct); wintering in the Sahel ✦ Fairly common (hundreds of thousands of pairs). Frequent on *páramos* in N; mainly restricted to mountains in S. Nests in Balearics. Scarce and irregular in Canary Is.

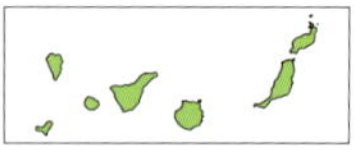

Berthelot's Pipit *Anthus berthelotii* (bisbita caminero). ~14 cm. Canary Is, where very common and the sole nesting pipit (also present in Salvages and Madeira). Similar in size to Meadow and Tree Pipits and also highly streaked; however, greyer, with pale legs and posture recalling Tawny Pipit. Bill fine, blackish with pale base. Song similar to Tawny Pipit: *cheer–lee*, repeated monotonously in flight ✦ All open habitats, or those with few trees, from sea level to >3.000 m on Mt Teide ✦ Sedentary ✦ Very common (tens of thousands) and widespread throughout all islands; **DD**

Richard's Pipit *Anthus richardi* (bisbita de Richard). ~18 cm. Rare. Like a juvenile Tawny Pipit lacking loral stripe and with malar stripe widening towards the collar. Robust, long-legged, frequently stretching upright. Bill stout; supercilium broad; tail long, dark, white-edged; breast and flanks warm buff; hindclaw long and straight. Flight call: a sparrow-like *chrrreep* ✦ Level areas with short grass, generally on coasts ✦ Wintering (Oct–Apr) and passage; breeds onwards from S Siberia ✦ Reported especially from Asturias and Galicia, but also from E and S coasts. Some records from Balearics. Vagrant to Canary Is.

White-winged Snowfinch
juv.
ad.
Tawny Pipit
ad.
Berthelot's Pipit
Richard's Pipit
ad.
juv.

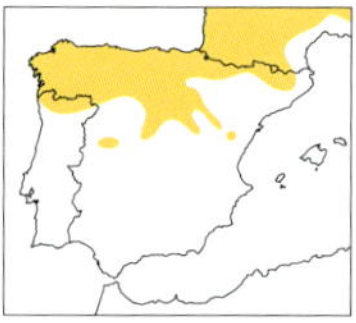

Tree Pipit *Anthus trivialis* (bisbita arbóreo). ~15 cm. Summer visitor, northern. Call: a loud, sharp, somewhat buzzing *zeep* or *speez*. Slightly larger and less greenish than Meadow Pipit, throat and breast tinged buffish cream; streaked strongly on breast but only thinly on flanks. Bill largish; head markings and median coverts more contrasting than in Meadow Pipit; legs pink, hindclaw curved, short. Often perches on trees and shrubs. Song spectacular, loud, with a slow final *chew–chew–chew*..., from a tree or in parachuting song flight ✦ Forest clearing and margins (oak and pinewoods) and plantations; also meadows, pastures and open scrubland with trees. Often well up mountains ✦ In summer and on passage (late Mar–early May and Aug–Oct) ✦ Fairly common (a few hundred thousand), although mostly confined to N half; apparently spreading. Common on passage in Balearics and Canary Is.

Meadow Pipit *Anthus pratensis* (bisbita pratense). ~15 cm. In winter. Very similar to Tree Pipit but call when taking flight distinct: a repeated *seep seep seep*. Relatively compact, brownish: with greenish tints in autumn plumage, highly streaked above and on breast and flanks. Hindclaw straight, long. In flocks. Runs about on ground between vegetation ✦ Diverse open terrain: croplands, pastures, fallows, marismas ✦ Wintering and passage (Sept–Nov and Mar–Apr) of birds from N and C Europe ✦ Abundant (millions) and widespread in winter, commonest on extensive cultivated plains and around wetlands. Passage conspicuous. Nests sporadically in Cantabrian range, on grassy hillsides with gorse or heather. Scarce but regular in Canary Is.

Water Pipit *Anthus spinoletta* (bisbita alpino). ~17 cm. Large, slim, dark-legged. In summer has diagnostic, pink-tinted, plain breast, brightest in male; grey head with white supercilium. Bill long, blackish; two wingbars; long tail with white outer rectrices. More like other pipits in winter, with breast streaked, malar stripe and paler bill and legs. Call: a brief, sharp *tseep*. Song, from perch or in parachuting song flight, a striking accelerating repetition of metallic notes ✦ Montane pastures at mid or upper levels, frequently at 1,500–2,500 m and near water (streams, glacial lakes, bogs). Winters around wetlands; rivers, pools, reservoirs, ricefields ✦ Absent from mountains from Oct/Nov to Mar/Apr. Wintering and passage of Alpine birds ✦ Common or very common (tens of thousands of pairs) in its localised breeding habitat in mountains of N and C; scarce or uncommon, but widespread, in winter. Vagrant to Canary Is.

Rock Pipit *Anthus petrosus* (bisbita costero). ~17 cm. Northern coastlines in winter. Slim and dark-legged, like Water Pipit, but darker and more heavily streaked, with barely any supercilium and greyish outer rectrices. Often solitary ✦ Coasts ✦ Wintering (Oct–Mar) and sparse passage of Scandinavian birds (ssp. *littoralis*) ✦ Scarce and scattered in N; occasional or vagrant elsewhere. Vagrant to Balearic and Canary Is.

Red-throated Pipit *Anthus cervinus* (bisbita gorjirrojo). ~15 cm. Rare migrant. Reddish face and throat conspicuous in spring, especially in male. In autumn very heavily streaked, especially on flanks (where streaks comprise two or three long broad lines) and mantle; rump also streaked. Hindclaw long. Call: a long, metallic, penetrating *pseeeee* or *skeeeee* ✦ Grassland with pools, often at coastal wetlands ✦ On passage (Apr–May and Sept–Oct), nesting onwards from N Scandinavia and N Russia ✦ Scarce but regular in spring, rare in autumn (but perhaps overlooked) In mediterranean coastlands (Catalonia, Valencia, Balearics, Strait); occasional or vagrant elsewhere in Peninsula. Scarce and irregular in Canary Is.

Tree Pipit
Meadow Pipit
Water Pipit
winter
summer
Rock Pipit
winter
summer
Red-throated Pipit

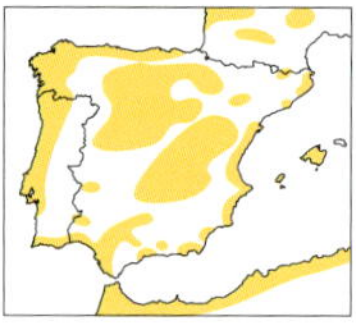

Yellow Wagtail *Motacilla flava* (lavandera boyera). ~17 cm. More compact and shorter-tailed than other wagtails. Adults yellow below, as Grey Wagtail, but greenish on mantle and rump. Juvenile different: brownish or greyish above and whitish or creamy below, with indistinct malar stripe that extends towards breast sides (recalling juvenile White Wagtail). **Citrine Wagtail** *Motacilla citreola* (lavandera cetrina), occasional (p. 237) is greyish above with two, broad, white wingbars. Yellow Wagtail races distinguishable in adult males (see Plate) but intergrades and indeterminate individuals are frequent. Call distinctive: a sharp, pleasant *tsweep* or *tsi-weep*. Song comprises short, shrill trills. Roosts in reedbeds ✦ Level damp grassy terrain: meadows, sedgebeds, pool margins, marismas, irrigated croplands, saltpans ✦ In summer and on passage (Mar/May–Sept/Oct), with a few overwinterers. Winter quarters in tropical Africa ✦ Ssp. *iberiae*: breeds quite commonly (hundreds of thousands of pairs) but only frequent in N half (Duero and Ebro basins); localised around wetlands in E and S. Sizable passage (ssp. *flava*, of France and C Europe with, in lesser numbers, ssp. *flavissima*, mainly of British Isles, some *cinereocapilla*, of Italy, and *thunbergi*, of Scandinavia; also vagrant *feldegg*, of the Balkans). Common migrant in Canary Is.

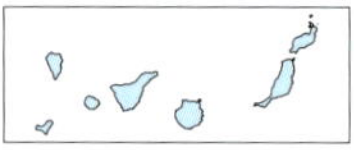

White Wagtail *Motacilla alba* (lavandera blanca). ~18 cm. Commonest wagtail. Dark cowl, constant tail pumping and head jerking when walking are distinctive. Black of breastband extends to throat in summer. Forehead white; cap centre black, grey in winter female; nape black in male, grey in female. Juvenile greyish and more uniform. **Pied Wagtail** Ssp. *yarrellii* (lavandera blanca enlutada) has back black and flanks dark grey. Flight call: a loud *chi-sic* (quieter than Grey Wagtail). Song: a warbling, swallow-like twitter. Flight undulating, with wings closed between bouts of rapid wingbeats. Follows plough. Forms large winter roosts, often in trees in town centres ✦ Chiefly alongside rivers, streams, irrigation channels near or within villages and homesteads (where nests). Often feeds in young crops, fallows, ricefields, marshes, pastures ✦ Spanish birds largely sedentary, with altitudinal movements; passage and wintering of European birds (Oct/Nov–Mar/Apr) ✦ Ssp. *alba*: nests commonly (hundreds of thousands of pairs), very common in N but progressively scarcer and more local towards S, where chiefly montane. Abundant in winter (millions) and very widespread, especially in S. Ssp. *yarrellii* (British Isles): only on passage and in winter, chiefly in N and W where scarce or uncommon.

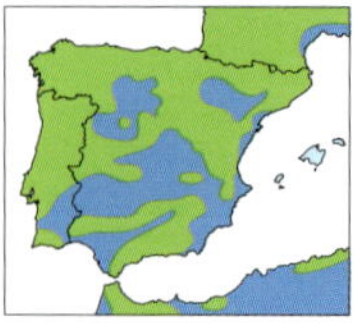

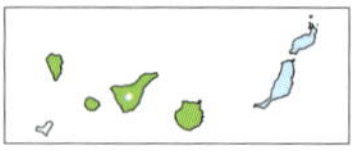

Grey Wagtail *Motacilla cinerea* (lavandera cascadeña). ~19 cm. Longest-tailed and slimmest wagtail. Grey above, rump and underparts – at least the vent – yellow. Pumps tail constantly and energetically. Shows white wingbar in flight. Throat black in male, white or grey in female, in summer. Call harsher than White Wagtail: *tseep* or *tsee–seep*. Song: a short series of notes similar to the call ending in a trill ✦ Rivers and streams, chiefly fast-flowing, most often in mountains, where sometimes at considerable altitude. Nests between rocks or in holes in bridges, dams, watermills. More widespread in winter, when often in cities ✦ Spanish birds largely sedentary; some winterers from C and N Europe (Sept/Nov–Feb/Apr) ✦ Ssp. *cinerea*: uncommon nester (tens of thousands of pairs), most frequent in N; moderate numbers of winterers; nests in Ceuta; scarce in Balearics and eastern Canary Is in winter. Ssp. *canariensis*: nests in C and W Canary Is (except El Hierro), common and well distributed, but declining; **DD**

ssp. iberiae
♀
ssp. flava
♂
ad. summer
ssp.
flavissima
Yellow Wagtail
ssp.
cinereocapilla
ssp.
thunbergi
White Wagtail
winter
ssp. feldegg
summer
♀
Grey Wagtail
♂ summer

FINCHES (Passeriformes, Fringillidae) – Typical seed-eaters, with strong, conical, frequently stout bills. Have only 9 primaries. Gregarious. Widespread worldwide.

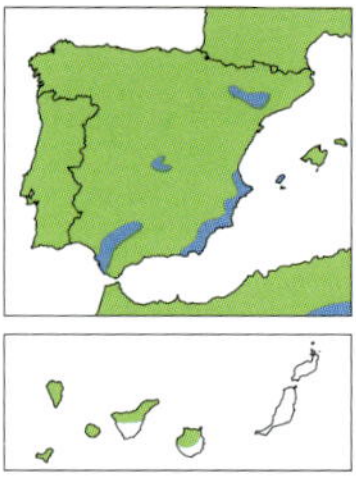

Common Chaffinch *Fringilla coelebs* (pinzón vulgar). ~15 cm. Very common. Sparrow-sized, slim; double wing bar and white outer rectrices very obvious in flight. Male colourful, less so in fresh autumn plumage; differs from those in Canary Is and N Africa (see Plate). Female and juvenile brownish, with pale nape patch. Calls: a loud *pink*, *pink*, recalling Great Tit; *tsip* or *chip* in flight. Song, repeated at short intervals, is a brief, loud, accelerating sequence ending in a flourished *tsweeoo*. Flight strong, undulating. Tends to feed on ground. Gregarious outside breeding season, often associating with other finches ✦ All forest types and any areas with sizable trees: riparian woods, dehesas, olive groves, orchards, parks. Very widespread in winter, when frequently on fallows and stubbles. In Canary Is, chiefly in laurel forest, tree heathland and mixed pinewoods ✦ Partial migrant, with wintering (Sept/Nov–Feb/Apr) of European birds ✦ Ssp. *coelebs*: very common breeder (several million pairs) and widespread, although scarce or absent in arid, treeless regions; increasing. Common in Mallorca and Menorca. Abundant (many millions) and very widespread in winter. Ssp. *africana* (NW Africa): resident in Ceuta (abundant) and Melilla; has reached the Peninsula on rare occasions. Generally common or very common in Canary Is although strictly tied to woodlands and tree heath shrubland, with the following endemic subspecies: *canariensis* (=*tintillon*) in Gran Canaria, Tenerife and La Gomera, **DD**; *palmae* in La Palma, **EN**, and *ombriosa* in El Hierro, **EN**

Blue Chaffinch *Fringilla teydea* (pinzón azul). ~17 cm. Canarian endemic, restricted to Tenerife and Gran Canaria. Large, quite uniformly coloured; male blue-grey, more intensely blue on head and mantle; female and juvenile brownish-grey. Belly pale. Double wingbar but much less evident than in Common Chaffinch and scarcely apparent in Tenerife males; lacks white in tail. Head and bill more robust than in Common Chaffinch; silhouette similar, with peaked crown. Habitual call: a disyllabic *cheeoo–cheeoo* or *wee-cheeoo*; song recalls Common Chaffinch but shorter and weaker. Usually solitary or in small flocks ✦ Chiefly pinewoods (Canarian and Monterrey pines), particularly at ~1,000–2,000 m ✦ Sedentary, with occasional altitudinal displacement by snow ✦ Ssp. *teydea*: Tenerife, relatively common (hundreds of pairs) in the pinewoods around Mt Teide (Vilaflor, Aguamansa), where favoured by forestry plantations; **VU**. Ssp. *polatzeki*: Gran Canaria, scarce (a couple of hundred) and local (Inagua, Ojeda and Pajonales pinewoods); **CR**

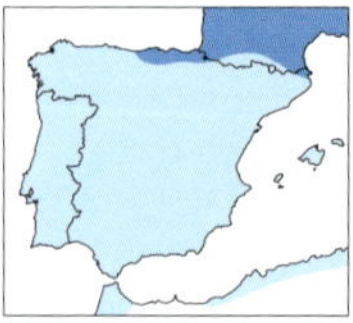

Brambling *Fringilla montifringilla* (pinzón real). ~15 cm. Only in winter. Similar in size and shape to Common Chaffinch but readily distinguished by its white rump, nearly all black tail and limited white in wings. Orange-toned on breast, 'shoulders' and lesser coverts. Sexes more similar in winter: male has head and mantle spotted black (these wearing to all black in spring) and yellow, black-tipped bill; female shows much grey on sides of head and neck, bill duller. Usual calls: a loud *chewik* and *chec, chec*… in flight ✦ Breeds in the taiga, wintering in woods (especially beechwoods) and croplands (including maize and sunflower stubbles) ✦ Wintering (~Oct–Mar, especially Dec–Feb) and passage of N European birds, some of which reach the extreme N of the Maghreb ✦ Relatively scarce and irregular, commoner in the N and in cold winters. Scarce in Balearics. Vagrant to Canary Is.

♀ ad.
Common Chaffinch
♂ ad. summer
♂ ad.
ssp. *canariensis*
Blue Chaffinch
♂ ad. summer
♀ ad.
♂ ad. autumn
Brambling

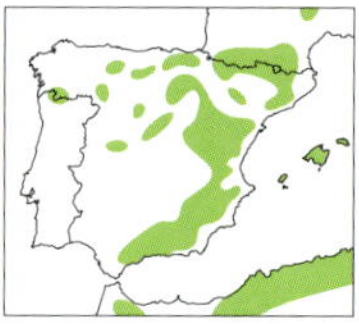

Red Crossbill *Loxia curvirostra* (piquituerto común). ~16 cm. Pinewoods. Sturdy, bill with crossed mandibles looks parrot-like at a distance. Plumage fairly uniform in adults: red in male (orange in immature male) and olive green in female. Juvenile, greyish brown, very streaked. Call distinctive: a loud, metallic *chip–chip* or *glip–glip*. Song Greenfinch-type but slower and faltering, with interspersed call notes. Flight strong. Gregarious year-round. Extracts pine seeds, sometimes adopting acrobatic postures ✦ Conifer woods and plantations. Nests generally in mountains but visits lowland pine woods, parks, etc. at other times ✦ Sedentary or nomadic, nesting very early in the year and dispersing immediately afterwards; irruptions of birds from N and C Europe frequent ✦ Ssp. *curvirostra*: uncommon nester (a few hundred thousand) with inter-annual fluctuations. Commoner in E half; more widespread in winter. Ssp. *balearica*: very common in Mallorca and Ibiza. Ssp. *poliogyna*: in Melilla. Possible vagrant to Canary Is.

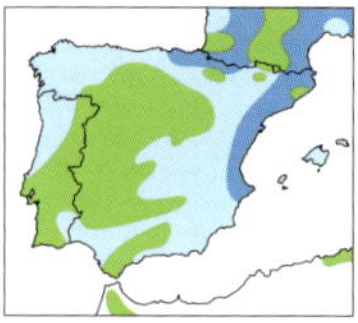

Hawfinch *Coccothraustes coccothraustes* (picogordo común). ~18 cm. Uncommon. Large, compact, bill enormous, head stout and tail short. Shows broad white band across primaries in flight. Call: a loud, explosive, metallic *pik* or *chik*. Bill grey in spring, yellowish in winter; bib and lores black, pale wingbar also across upperwing coverts; rectrices tipped white. Female duller, with grey panel on secondaries. Juvenile lacks black bib, spotted black below. Flight strong, undulating. Song short and weak. Secretive, often overlooked ✦ Woods and copses (deciduous oaks; Portuguese, Holm and Cork oaks; beeches, nettle trees… sometimes pines) or riverine woods and parks, these chiefly in winter ✦ Partial migrant, Spanish birds sedentary and irregular wintering of C European birds (~Oct–Mar/Apr) ✦ Scarce breeder (thousands of pairs) although perhaps well distributed, most frequent in C and SW. More widespread in winter, when frequent in Catalonia and Valencia, and scarce in Balearics. Vagrant to Canary Is.

Eurasian Bullfinch *Pyrrhula pyrrhula* (camachuelo común). ~16 cm. Northern. Male unmistakable. Female duller, with greyish tones above and pinkish below. Juvenile lacks cap. Rump, undertail and secondary wingbar white; tail and wings blackish; bill short and stout. Call: a plaintive piping *peeoo*. Song quiet, a variable mixture of piping and squeaky notes. Solitary, in pairs or in small flocks. Shy. Often feeds on buds on twigs of trees and bushes ✦ Humid woodlands, mainly in bushy clearings and margins; also in open countryside with trees. More widespread in winter, visits riparian woods and fruit orchards ✦ Partial migrant, Spanish birds mainly sedentary with small-scale wintering of European birds (ssp. *europoea*) ✦ Ssp. *iberiae*: uncommon breeder (hundreds of thousands) although widespread in its range. Scarce and scattered elsewhere in the Peninsula in winter.

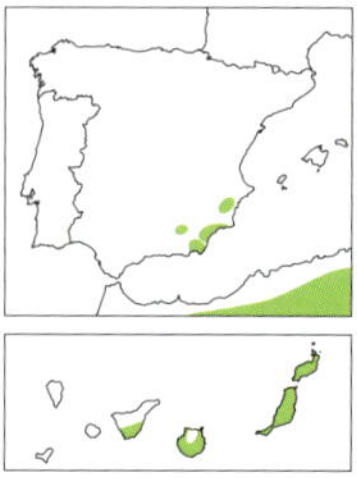

Trumpeter Finch *Bucanetes githagineus* (camachuelo trompetero). ~12 cm. Arid SE and Canary Is. Small, stocky, sandy-toned, with short, stout, yellowish or reddish bill. Calls nasal, metallic, like a toy trumpet. Breeding male has bill and forehead red; head grey; rose-tinted on face, rump, wings, tail and belly. Female duller. Juvenile grey-brown. In small groups ✦ Broken, semidesert terrain, with rocks, escarpments, dry river beds ✦ Resident and dispersive ✦ Ssp. *zedlitzi*: recent colonist (1970s), still scarce (a few hundred pairs), distributed between sites in Almería, Granada, Murcia and Alicante; vagrant elsewhere, including the Balearics; **NT**. Ssp. *amantum*: Canary Is, fairly common (some thousands), declining; frequent in Lanzarote and Fuerteventura; **EN**

♀ ad.
Red Crossbill
♂ ad.
Hawfinch
♀ ad.
♂ ad. summer
♀ ad.
♂ ad.
♂ ad. summer
Eurasian Bullfinch
Trumpeter Finch

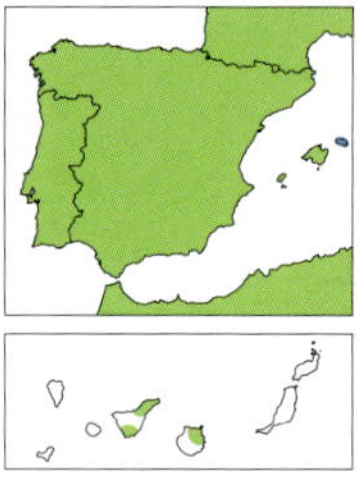

European Serin *Serinus serinus* (serín verdecillo). ~11 cm. Common. Small, greenish, rump yellow. Bill small. Male also bright yellow on forehead, supercilium, sides of neck, throat and breast. Two short, narrow wingbars; tail plain dark; heavily streaked above and below, including breast in female. Juvenile light brown, lacks yellow, highly streaked. Call: a buzzy *trililit* in flight; *tsweee* in alarm. Song, perched or in striking, slow, display flight, a rapid mix of melodious, high-pitched, jangling notes and short buzzy trills. Flight rapid, undulating. Highly gregarious ✦ Woodland clearings and edges, and all types of areas with trees, riparian woods, hedges, gardens, plantations and orchards. Also in open country in winter ✦ Partial migrant, Spanish birds largely migratory and with passage and wintering of European birds. Passage (~Sept–Nov and Feb–Apr) ✦ Abundant (millions of pairs) and widespread, most frequent in Mediterranean habitats. Established locally in Gran Canaria and Tenerife (since 1970s).

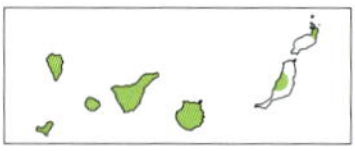

Island Canary *Serinus canaria* (serín canario). ~13 cm. Endemic to Macaronesia. Ancestor of domestic canaries. Larger and more slender than Serin, with tail longer and wings shorter; bill larger; head greyer; less heavily streaked; male also yellow on flanks and belly; rump dull greenish-yellow, less conspicuous. Unlike Greenfinch tail and remiges unmarked. Calls: a sharp *cheweet*, *chiodididi* in flight. Song varied and musical (as familiar cagebird) ✦ Croplands, woodland edges, escarpments, montane scrub, pastures with thistles, gardens ✦ Sedentary ✦ Abundant (tens of thousands of pairs) and widespread in the central and western islands; local in the east; many are captured. **DD**

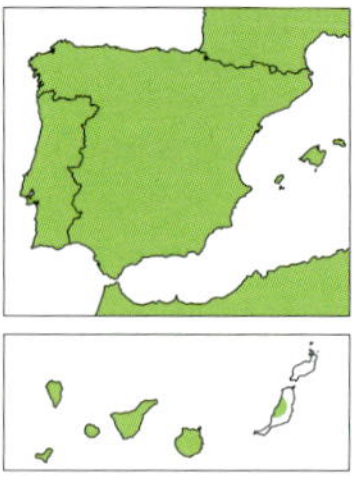

European Greenfinch *Chloris chloris* (verderón común). ~15 cm. Larger than other greenish finches, with striking yellow markings on primaries and tail sides. Size and large head and bill as House Sparrow but tail forked and flight undulating. Male more yellow-tinged, especially in spring. Juvenile browner, streaked. Calls: a trilling *chil–il–il–il*; *chip*, *chewit*. Song, often in flight, a mix of Canary-like trills with a peculiar nasal, stuttering, often repeated phrase: *zweeeer* or *chrrweee*. Often in large flocks ✦ Open countryside with trees, orchards, riparian trees, gardens, etc., also open woodlands, frequently Mediterranean ✦ Partial migrant, with notable passage and wintering (~Oct/Nov–Mar/Apr) of C European birds (ssp. *chloris*) ✦ Ssps. *vanmarli* (NW, Portugal, Ceuta and Melilla) and *aurantiiventris* (elsewhere in Peninsula and Balearics): very common (several millions of pairs) and widespread, most abundant in warmer regions; increasing. Common and spreading in Canary Is (since 1970s).

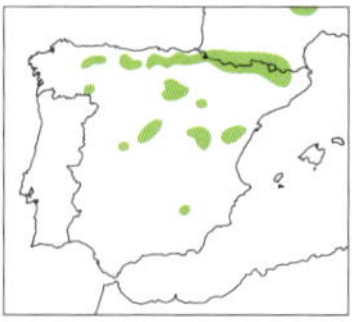

Citril Finch *Carduelis citrinella* (verderón serrano). ~12 cm. Mountains of C and N. Greenish, with nape and necksides grey. Differs from Serin in having two broad yellow wingbars; from Greenfinch and Siskin in plain tail. Bill intermediate between Serin and Siskin. Male unstreaked, face very yellow; female streaked above; juvenile brown, streaked above and below, with buffish-rufous wingbars. Flight call distinctive: a metallic, somewhat nasal *ti–ti–ti–ti*. Song, often in fight, has Serin-like trills and musical, Goldfinch-like twittering ✦ Woodland clearings and edges, chiefly of pines, sometimes, beeches; generally above 1,000 m, often at the treeline ✦ Sedentary, with altitudinal and dispersive movements ✦ Common in its nesting areas (tens of thousands of pairs); spreading. Vagrant to Balearic Is.

♂ ad.
European Serin
♀ ad.
♂ ad.
Island Canary
♀ ad.
♀ ad.
European Greenfinch
♂ ad.
♀ ad.
Citril Finch
♂ ad.

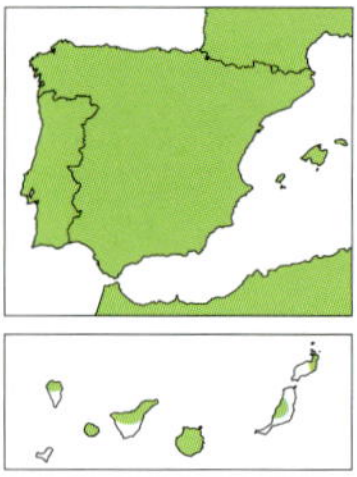

European Goldfinch *Carduelis carduelis* (jilguero europeo). ~12 cm. Common and familiar. Black wings with obvious broad yellow wingbars; rump white, tail black with white markings. Tinkling calls distinctive, liquid and musical: *tuu-wee-oo, tee–lee…*; song pleasant, including frequent call-type notes. Sexes similar. Adults have distinctive red face mask and black and white head markings; juvenile has head pale, finely streaked. Bill ivory-white, fine and pointed. Flight markedly undulating. Gregarious, often in mixed-species flocks ✦ Riparian trees, open farmland with trees, orchards, rough pastures with thistles, village outskirts, less often in woodland clearings. Eats seeds of thistles and other weeds ✦ Partial migrant, some Spanish birds wintering in N Africa and large-scale wintering of birds from France, Germany, British Isles, etc. (ssps. *carduelis* and *britannica*). Passage around Oct–Nov and Mar–Apr ✦ Ssp. *parva*: very common breeder (millions of pairs) and widespread, most frequent in low areas with trees. Abundant in winter. Many captured for cagebird trade. Uncommon in Canary Is.

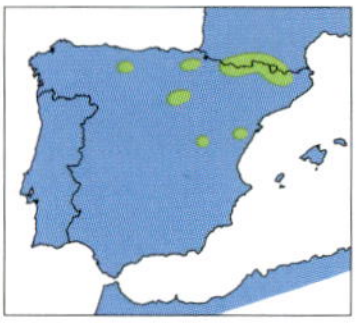

Eurasian Siskin *Spinus spinus* (jilguero lúgano). ~12 cm. Small, and greenish-yellow like Serin and Citril Finch but uniquely combines yellow wingbars, rump and tail-sides. Male has cap and throat black. Female and juvenile have streaked cap; juvenile wingbars whitish. Bill fine, Goldfinch-type; tail short; flanks streaked. Calls short and sharp *toolee*, *tsuee*… Song, often in flight, a rapid jumble high-pitched jangling, including metallic twitters, trills and wheezy notes, mixed with nasal, Greenfinch-like, harsher notes and mimicry. Highly gregarious. Feeds mainly in trees, often hanging acrobatically from cones and catkins ✦ In winter frequents alders and birchwoods, whose seeds it eats; also pinewoods, eucalyptus plantations, open country with trees. All nests in Spain are in montane conifers ✦ Chiefly in winter (~Oct/Nov–Mar/Apr) and on passage to N Africa, originating from N and C Europe; some remain to breed ✦ Winters in very variable numbers, with occasional massive irruptions. Widespread but most frequent in littoral regions of E (Catalonia, Valencia) and N (Basque Country). Breeding sparse and sporadic (hundreds of pairs, more after irruptions), in mountain forests, mainly in the Pyrenees. Common in winter in the Balearics (has bred). Occasional in the Canary Is.

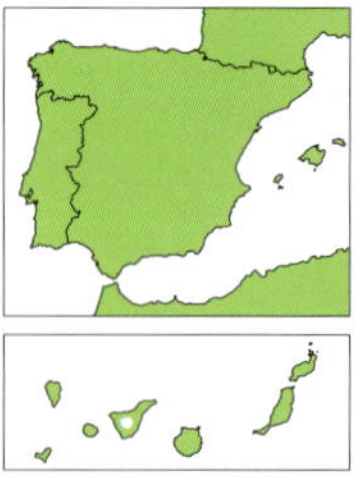

Common Linnet *Linaria cannabina* (pardillo común). ~13 cm. Very common in open country. Brownish, with whitish primary edges and tail sides. Sexes similar in autumn. Wear gradually reveals the showy red breast patches and forehead of the adult male; head otherwise grey and back chestnut. Juvenile resembles female, but duller and more streaked. Bill short, dark. Flight call: a brief trilling *ti–ti–tirit* or *chichichichit*; alarm: *tsweet*. Song musical and pleasant, though weak, a rapid mix of trills, warbles and whistles; sometimes with female participation. Highly gregarious. Nests often very close together ✦ Nests in all types of scrub (gorse, heather, box, juniper, cistus, broom, kermes oak…) and in shrubs, hedges, vineyards). Forages on open hillsides, pastures, fallows, stubbles, croplands; at very variable elevations ✦ Partial migrant, Spanish birds wandering or migratory, and numerous wintering birds from C and N Europe (ssp. *cannabina*); passage around Oct–Nov and Mar–Apr ✦ Ssp. *mediterranea*: very common (millions of pairs) and very widespread, most frequent in warmer regions. Wintering and passage notable in E. Captured for cagebird trade. Canary Is; endemic ssps. *harterti* (Lanzarote and Fuerteventura) and *meadewaldoi* (elsewhere): widespread and variably common; **DD**

juv.
♀
♂
ad.
European Goldfinch
♀ ad.
Eurasian Siskin
♂ ad.
Common Linnet
♂ ad.
♀ ad.

BUNTINGS (Passeriformes, Calcariidae –Snow Bunting– and Emberizidae –others) – Nine-primaried seed-eaters, like finches. Bill conical and broad, somewhat swollen. Select herbaceous seeds. Eurasia and Africa (Emberizidae) and N America (most Calcariidae).

Snow Bunting *Plectrophenax nivalis* (escribano nival). ~17 cm. Rare winterer on northern coasts. Large white wing and tail patches recall Snowfinch (p. 206) but head brownish-yellow, not grey, and distributions separate. Sexes quite similar in winter plumage but female shows less white on wing, especially in 1st winter. Flight call distinctive: a buzzing *tirrirrirrip* or *tirrirrilit*; also a plaintive *pyew*. Wings long; tail notched, bill yellowish with black tip; whitish below, with buff breast sides and flanks. Highly terrestrial. Alone or in small flocks ✦ Usually on beaches, sand-dunes and other open areas ✦ In winter (Oct–Mar, mainly Nov–Feb), with northern breeding areas including N Scotland, Scandinavia and Iceland ✦ Rare and irregular in Galicia and the Biscay coastlands; occasional in Catalonia; Vagrant to other regions, Balearics and Canary Is.

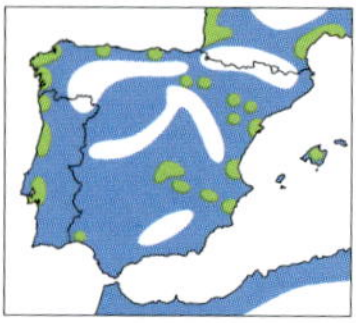

Reed Bunting *Schoeniclus schoeniclus* (escribano palustre). ~15 cm. Common in winter. Male distinctive in spring, with head and bib black and broad white collar and moustachial stripe; these are obscured by pale feather tips in fresh plumage in autumn. Female brown-headed with creamy supercilium; dark cheeks and ear coverts, cream submoustachial stripe and broad black malar stripe. Very streaked above, sparrow-style; wings rufous-toned; flanks streaked; outer rectrices white, and obvious as tail is often flicked open. Bill short, larger in resident races. Call: *tsee* or *tseu*. Song: a few short notes followed by a rapid trill. Small winter flocks gather to roost in reedbeds ✦ Nests in marshland vegetation (reeds, reedmace, sedge clumps…); visits croplands, particularly in winter, frequent in irrigated valleys ✦ Partial migrant; Spanish birds sedentary with large-scale wintering (mainly Oct–Mar) of birds from C and N Europe ✦ Ssps. *lusitanica* (Galicia and N Portugal) and *witherbyi* (elsewhere in Peninsula and Balearics): scarce nesters (some hundreds of pairs) and highly localised in coastal NW, La Mancha lagoons, Ebro delta and Mallorca; **EN**. Ssp. *schoeniclus*: widespread and common in winter, most frequent in E half, but strongly declining; **VU**. Vagrant to Canary Is.

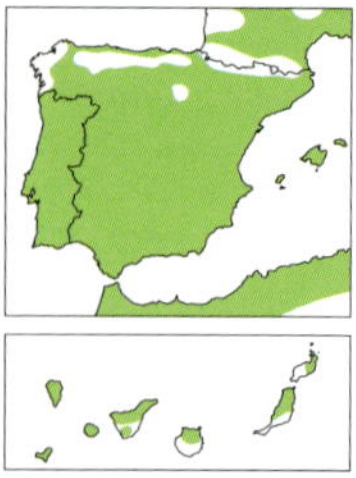

Corn Bunting *Emberiza calandra* (escribano triguero). ~18 cm. Abundant. Brownish and streaked, recalling a lark or female sparrow, but sizable with heavy bill. Lacks distinctive fieldmarks other than dark malar stripe that tends to merge with breast streaks. Sexes similar but male bulkier. Flight call: a clicking *tic* or *pit*, sometimes *pit–it–it*. Characteristic song, often repeated and at any hour, has a halting start and soon accelerates to a jangling ending: *chic–chic–chi–chi–chirrirríííí*; usually given from an exposed songpost (wire, post, fence). Courting male often flies with legs dangling. Often in flocks ✦ All types of open country at lower altitudes, chiefly cereal croplands and grasslands; also dehesas, pastures, olive groves, irrigated crops, scrub margins and clearings, etc. ✦ Partial migrant, Spanish birds variably migratory, with some winterers from C Europe. Passage around Sept–Nov and Feb–Apr ✦ Very common (millions of pairs) and widespread throughout all Mediterranean landscapes that are not too arid, but rare in mountains and in Galicia and the Biscay coastlands. Very common and widespread in the Balearics. locally common in the Canary Is (scarce in Lanzarote and Fuerteventura).

♂ ad. summer
juv.
Snow Bunting
♀ ad.
juv.
♀ ad.
Reed Bunting
♂ ad. summer
ssp. *witherbyi*
Corn
Bunting

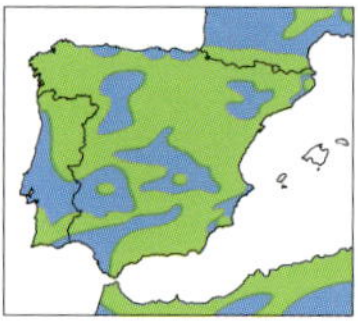

Rock Bunting *Emberiza cia* (escribano montesino). ~16 cm. Hilly terrain. Body orange-brown; mantle streaked blackish; head grey, striped black; in flight shows rufous rump and white outer rectrices. Male brighter and more contrasted than female. Juvenile duller, with brownish head and streaked throat, breast and flanks. Calls weak, sharp: *chi*, *sit*. Song longer and more varied than other buntings, with some trills (recalls Dunnock) ✦ Steep, open, dry, stony, mountain slopes and hillsides at variable altitudes, or clearings in hillside woodlands or scrub. Visits croplands, especially in winter ✦ Sedentary, with altitudinal movements ✦ Breeds commonly (a few million pairs) and widespread, most abundant in Mediterranean areas that are not too dry. Occasional in Balearics in winter.

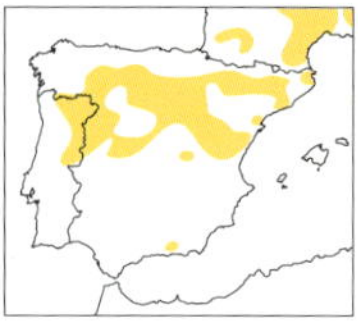

Ortolan Bunting *Emberiza hortulana* (escribano hortelano). ~16 cm. Summer visitor. Orange-brown below like Rock Bunting, with yellow throat and submoustachial stripe, pinkish bill and legs and striking pale eye-ring. Streaked above and on brown rump. Male has head and breast greenish, unstreaked. Female has dark malar stripe and streaked breast. Juvenile darker and more streaked below. Calls include a ringing disyllabic *tsleeu* and a sharp *quip*. Brief song is several ringing repetitions of one note with a lower-pitched ending, sometimes omitted: *chi–chi–chi–chi–du–du* ✦ Sunny mountainsides and other open, sometimes dry, shrubby areas. At considerable altitude in C and S ✦ Trans-Saharan summer visitor (an exception among granivores). Passage mainly ~Apr–May and late Aug–Oct ✦ Uncommon (a few hundred thousand pairs), declining; unevenly distributed: most common on N Meseta and central mountains, only on the Sierra Nevada in the S. Movements poorly marked. Scarce migrant in Balearics. Vagrant to Canary Is.

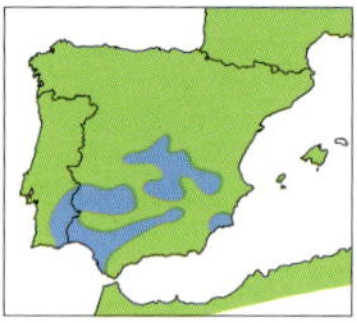

Cirl Bunting *Emberiza cirlus* (escribano soteño). ~16 cm. Widespread. Yellow-toned, like Yellowhammer, but rump plain olive- or brownish-grey. Male has distinctive facial pattern, less evident in fresh autumn plumage, and broad greenish breastband. Female has pale throat and streaked breast. Juvenile duller than female, less yellowish. Call: a gentle *tsii* or *tsip*. Song brief, monotonous, a rapid rattling on one note: *sre–sre–sre–sre–sre...* (recalls distant Bonelli's Warbler but louder, longer and metallic). Often in small flocks ✦ Open terrain with trees or tall shrubs, in countryside, woodland clearings, dehesas, hedges, orchards ✦ Resident, with short-distance winter movements ✦ Common (hundreds of thousands of pairs) and widespread, but most frequent in N and absent from treeless areas of C and S. Common in Mallorca and Ibiza. Possible vagrant to Canary Is.

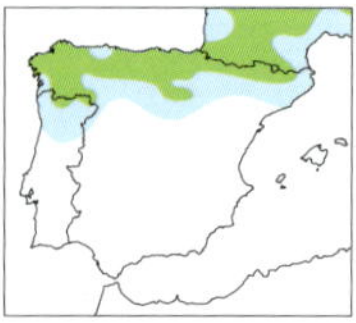

Yellowhammer *Emberiza citrinella* (escribano cerillo). ~16 cm. Northern. Yellow-toned, with unstreaked rufous rump and white outer rectrices. Breeding male has striking yellow head with some dark lines, rufous brown on chest and yellow belly. Female, especially 1st winter, duller, with darker head and streaked breast and flanks. Call: a sharp *tsit*. Song distinctive: a rapid succession of similar notes followed by a brief pause and ending in a drawn-out note: *si–si–si–si–si–Si...siiii*; given from a small tree, shrub or wire. Gregarious when not breeding ✦ Open countrysde with trees, hedges with trees, woodland and crop margins, open scrublands. Chiefly at 500–1,500 m ✦ Partial migrant, with Spanish birds resident or wandering and irregular wintering (mainly Nov–Feb), of European birds, particularly in cold winters ✦ Common (a few hundred thousand) but only in humid regions; declining. Wintering is largely within the breeding range. Vagrant to the Balearic Is.

Rock Bunting
♀ ad.
♂ ad.
♀ ad.
Ortolan Bunting
♂ ad.
♀ ad.
Cirl Bunting
♂ ad.
♀ ad.
Yellowhammer
♂ ad.

RARITIES

We here include only those species that occur occasionally or as vagrants in Spain and that:

a) have had records from 1984–2013 accepted by the Spanish rarities committee, CR-SEO (**Comité de Rarezas de la Sociedad Española de Ornitología)** as reported in the journal *Ardeola*,
b) have had at least one record accepted since 1 January 1950,
c) have apparently arrived naturally and not by escaping from captivity, deliberate release or accidental transportation.

In addition to genuinely wild species, comprising Category A of the Spanish bird list (*Lista de las Aves de España*, Gutiérrez *et al.*, 2012), there are some (Category C) that originate from long-established introduced or naturalised populations in other countries. A few species that have recently been added to the list of those considered by the Rarities Committee are not included here. They are the Bean Goose *Anser fabalis*, Corn Crake *Crex crex*, Small Buttonquail *Turnix sylvaticus*, Red-necked Phalarope *Phalaropus lobatus*, Roseate Tern *Sterna dougallii*, Lesser Crested Tern *Thalasseus bengalensis*, Great Grey Shrike *Lanius excubitor* and Lesser Whitethroat *Sylvia curruca*.

Following the English, scientific and Spanish names, we summarise the following:

- the broad breeding range of the species,
- the number of accepted Spanish records up to the end of 2013 (in bold),
- a summary of the distribution of the records according to the regions defined in the adjacent map and their periods of occurrence. Specific locations and years are given where there are only three records or fewer.

An asterisk (*) indicates former rarities that are no longer on the Rarities Committee list.

We emphasise the importance of collaborating with rarities committees, to ensure that valuable records form part of the scientific archive. In Spain, the CR-SEO is responsible for collecting, evaluating and publishing records of rarities that are properly documented (by photographs, drawings and/or descriptions). The postal address of the committee is C/. Melquiades Biencinto 34, 28053 Madrid, and email **rarezas@seo.org**. Website http://www.seo.org/conocenos/grupos-de-trabajo/comite-de-rarezas/; rarity records may be submitted in English using the bilingual (Spanish/English) submission form ('***esta ficha***') that is available on the website

Whooper Swan *Cygnus cygnus* (cisne cantor). Eurasia, boreal. **41** (24 Galicia, 11 North, 3 Levant, 1 Andalucía, 1 Catalonia and 1 Meseta; most Nov–Jan).

Tundra Swan *Cygnus columbianus* (cisne chico). Eurasia and N America, boreal. **8** (5 Catalonia, 1 Ebro, 1 Galicia and 1 Andalucía; Oct–Mar).

Red-breasted Goose *Branta ruficollis* (barnacla cuellirroja). Siberia; some may originate from captivity. **8** (6 Meseta, 1 Ebro and 1 Galicia; Nov–Feb).

Whooper Swan

Canada Goose *Branta canadensis* (barnacla canadiense grande). **C**. N America, introduced to British Is and Scandinavia; some may originate from captivity. **19** until 2005 (9 Meseta, 5 North; Sept–Mar). (*)

Bar-headed Goose *Anser indicus* (ánsar indio). C. Asia. >**40** up to 2010, most in wintering areas of Greylag Goose in Castilla and León and W Andalucía; ~Nov–Feb. (*)

Red-breasted Goose

Pink-footed Goose *Anser brachyrhynchus* (ánsar piquicorto). Greenland, Iceland and Spitsbergen. **65** (30 Meseta, 17 North, 11 Galicia, 5 Andalucía, 1 Catalonia and 1 Ebro; Sept–Apr, most Nov–Feb). (*)

Lesser White-fronted Goose *Anser erythropus* (ánsar chico). **C**. Eurasia, boreal, new arrivals seem to originate from reintroduction programmes in Scandinavia. **8** (5 Meseta, 1 Andalucía, 1 Balearics and 1 North; Oct–Mar).

King Eider *Somateria spectabilis* (éider real). Eurasia and N America, boreal. **5** (3 Galicia, 1 Catalonia and 1 Levant; Dec–Jun).

Siberian Scoter *Melanitta stejnegeri* (negrón siberiano). NE Asia. **1** (Galicia: Vigo and O Grove, Dec 11).

Canada Goose

Surf Scoter *Melanitta perspicillata* (negrón careto). N America. **20** (14 Galicia, 3 North and 3 Andalucía; Oct–Apr, most Dec–Jan).

White-winged Scoter *Melanitta deglandi* (negrón aliblanco). NW N America. **1** (Galicia: ría de Arousa, Dec 11).

Black Scoter *Melanitta americana* (negrón americano). NE Asia and northern N America, boreal. **3** (North: ría de Navia, Nov. 86; ría de Villaviciosa, Oct–Nov 94. Galicia: ría de Corme and Laxe, Feb–Mar 02).

Surf Scoter

Bufflehead *Bucephala albeola* (porrón albeola). N America. **3** (Galicia: laguna de Traba, Dec 92–Jan 93. North: Astillero, Nov 97–Jan 98 and Jan–Mar 10).

Barrow's Goldeneye *Bucephala islandica* (porrón islándico). N America and Iceland. **1** (Galicia: laguna de Traba, Jan–Mar 97).

Smew *Mergellus albellus* (serreta chica). Eurasia, boreal. **26** (12 North, 5 Galicia, 4 Catalonia, 3 Meseta and 2 Ebro; Aug–May, most Nov–Feb).

Hooded Merganser *Lophodytes cucullatus* (serreta capuchona). N America. **2** (Canaries: Tenerife, Dec 01–Feb 02. Meseta: Santa Cristina del Páramo, León, Dec 09).

Smew

Goosander *Mergus merganser* (serreta grande). Eurasia and N America, boreal. **53** (18 North, 12 Galicia, 9 Catalonia, 4 Ebro, 4 Levant, 4 Meseta and 2 Andalucía; Nov–Jun, most Dec–Jan). P. 24.

Ring-necked Duck *Aythya collaris* (porrón acollarado). N America. **142** (58 Canaries, 30 Galicia, 18 Meseta, 16 North, 10 Andalucía, 5 Catalonia, 3 Ebro, 2 Levant and 1 Balearics; Oct–May, most Nov–Mar). (*)

Lesser Scaup *Aythya affinis* (porrón bola). N America. **36** (17 Canaries, 13 Galicia, 2 Meseta, 2 North, 1 Andalucía and 1 Catalonia; Oct–Apr, most Nov–Mar).

Blue-winged Teal *Spatula discors* (cerceta aliazul). N America. **54** (33 Península, widespread, 4 Balearics and 17 Canaries; year-round, most Nov–Apr).

Baikal Teal *Sibironetta formosa* (cerceta del Baikal). Asia. **2** (Levant: Albufera de Valencia, Jan 74 and Jan 83).

Falcated Duck *Mareca falcata* (cerceta de alfanjes). Asia. **3** (Galicia: ría de Camariñas and laguna de Traba, Dec 92 to May 93. Galicia/Asturias: ría del Eo, Dec 99. Catalonia: delta del Llobregat, Jan–Mar 04).

American Wigeon *Mareca americana* (Silbón americano). N America. **44** (20 Galicia, 8 North, 5 Meseta and 11 Canaries; Sept–May, most Nov–Feb).

American Black Duck *Anas rubripes* (ánade sombrío). N America. **13** (11 Galicia plus North, 1 Catalonia, 1 Canaries; most Nov–Apr).

Pied-billed Grebe *Podilymbus podiceps* (zampullín picogrueso). Americas. **7** (4 Galicia, 1 North, 1 Ebro and 1 Canaries; scattered dates).

Lesser Flamingo *Phoeniconaias minor* (flamenco enano). Africa and India. **56** (25 Andalucía, 18 Meseta, 6 Catalonia, 5 Levant and 2 Canaries; year-round, peak Mar–Jun). P.36.

Red-billed Tropicbird *Phaethon aethereus* (rabijunco etéreo). Tropical oceans (nearest colonies in Cape Verdes). **30** (Canaries; spread throughout year; has bred).

Oriental Turtle-dove *Streptopelia orientalis* (tórtola oriental). Asia. **2** (Levant: Columbretes Is, Sept 01. Galicia: laguna de Traba, Oct–Nov 05).

Laughing Dove *Spilopelia senegalensis* (tórtola senegalesa). Africa and SW Asia. **17** in Península and Balearics (7 Andalucía, 4 Balearics, 3 Catalonia, 1 Galicia, 1 Levant and 1 Meseta; spread throughout year, peaks in Apr–May and Oct). Not regarded as a rarity in the Canary Is, where recently established as a breeding species in Fuerteventura and Lanzarote.

Common Nighthawk *Chordeiles minor* (añapero yanqui). N America. **2** (Canaries: Tenerife, Dec 98. Andalucía: Puerto Real, Nov 05).

Ring-necked Duck

Pied-billed Grebe

Red-billed Tropicbird

Laughing Dove

Chimney Swift *Chaetura pelagica* (vencejo de chimenea). N America. **6** (4 Galicia, 1 Andalucía and 1 Canaries; Oct–Nov).

Little Swift *Apus affinis* (vencejo moro). Africa, Middle East and India. **66** (48 Andalucía, 1 Catalonia, 1 Meseta and 16 Canaries; year-round, peaking in spring, around May, outside breeding areas in Andalucía). (*)

Yellow-billed Cuckoo *Coccyzus americanus* (cuclillo piquigualdo). N America. **1** (Balearics: Mallorca, Oct 94).

African Crake *Crex egregia* (guión africano). Africa. **5** (Canaries, Nov–Jan).

Sora Crake *Porzana carolina* (polluela sora). N America. **2** (Andalucía: Marismas del Guadalquivir, Dec 75. Galicia: Laxe, Apr 88).

Striped Crake *Amaurornis marginalis* (polluela culirroja). Africa. **1** (Andalucía: Las Navas de la Concepción, Dec 10).

Allen's Gallinule *Porphyrio alleni* (calamoncillo africano). Africa. **23** (9 Península, 2 Balearics, 1 Ceuta and 11 Canaries; 21 Nov–Mar, peak Dec).

Purple Gallinule *Porphyrio martinicus* (calamoncillo americano). N America. **2** (Canaries: Tenerife, Oct 92; Gran Canaria, Oct 04).

Lesser Moorhen *Gallinula angulata* (gallineta chica). Africa. **1** (Andalucía: Algeciras, Mar 03).

American Coot *Fulica americana* (focha americana). N America. **3** (North: Irún, Nov 99. Galicia: Sanxenxo, Jan–Mar 03; Valdoviño, Jan–Apr 09).

Sandhill Crane *Antigone canadensis* (grulla canadiense). N America. **2** (Extremadura: regadíos del Guadiana, Nov 11–Jan 12. Aragón: Gallocanta, Feb 13).

Demoiselle Crane *Anthropoides virgo* (grulla damisela). Asia. **3** (Extremadura: regadíos del Guadiana, Feb 66. Aragón: Gallocanta, Mar 66. Castilla and León: near Ataquines, Jan–Feb 04).

Pacific Loon/Diver *Gavia pacifica* (colimbo del Pacífico). N America. **1** (North: Santoña, Dec 09).

Black-bellied Storm-petrel *Fregetta tropica* (paíño ventrinegro). Southern Oceans. **2** (Canaries: Banco de la Concepción, Sept 11 and Aug 13).

Madeiran Storm-petrel *Hydrobates castro* (paíño de Madeira). Tropical and subtropical oceans (colonies in Azores, Madeira, Canaries and Berlengas, mainland Portugal). **12** in mainland Spain (10 Galicia, 1 Andalucía and 1 Levant; most Jul–Oct). P. 56.

Swinhoe's Storm-petrel *Hydrobates monorhis* (paíño de Swinhoe). E. Asian seas. **5** (3 Canaries and 2 Levant; Jun–Sept).

Chimney Swift

Sandhill Crane

Pacific Loon/Diver

Swinhoe's Storm-petrel

Black-browed Albatross *Thalassarche melanophris* (albatros ojeroso). Southern Ocean and Pacific. **6** (4 Galicia, 1 Asturias and 1 Andalucía; scattered through the year).

Cape Petrel *Daption capense* (petrel damero). Antarctic and sub-Antarctic isles. **1** (off Andalucía: in Gibraltar waters, Jun 79).

Black-capped Petrel *Pterodroma hasitata* (petrel antillano). Antilles. **1** (Galicia: 180 miles NW of Cape Finisterre, Apr 02).

Cape Verde/Desertas Petrel *Pterodroma feae/deserta* (petrel gongón/de las Desertas). **6** (5 Canaries, 1 Galicia; Jun–Dec). A further 11 records (7 Península, 4 Canaries) accepted as *Pterodroma* sp., also encompassing **Zino's Petrel** *P. madeira* (petrel freira).

Cape Verde Shearwater *Calonectris edwardsii* (pardela cenicienta caboverdiana). Cape Verde Is. **3** (Canaries: Montaña Clara islet, Lanzarote, 2 in May 10 and 1 in Jun–Jul 13).

Audubon's Shearwater *Puffinus lherminieri* (pardela de Audubon). Ssp. *baroli* (formerly included in *P. assimilis*). Tropical and subtropical Atlantic (colonies in Azores, Madeira and Canaries). **5** in mainland Spain (3 Andalucía and 2 North; diverse dates). P. 60.

Bulwer's Petrel *Bulweria bulwerii* (petrel de Bulwer). Tropical and subtropical waters of Atlantic and Pacific (colonies in Azores, Madeira and Canaries). **3** in mainland Spain (Andalucía: marismas del Odiel, Jun 00; Isla Cristina, Oct 06. Galicia: off Cape Finisterre, Aug 10).

Marabou *Leptoptilos crumenifer* (marabú africano). Africa. **27** (11 Andalucía, 5 Ebro, 5 Meseta, 3 Catalonia, 2 North and 1 Levant; year-round, peaking around Apr and Sept).

African Sacred Ibis *Threskiornis aethiopicus* (ibis sagrado). **C.** Africa, feral in France. **27** between 1994 and 2005 (widely dispersed in place and time). (*)

Northern Bald Ibis *Geronticus eremita* (ibis eremita). Morocco and perhaps still in Near East. **2** (Andalucía: Marismas del Guadalquivir, Jul 58. Meseta: Piedrahita/La Aldehuela, Ávila, Dec 04–Jan 05). Introduced in Cadiz, Andalucía.

American Bittern *Botaurus lentiginosus* (avetoro lentiginoso). N America. **5** (3 Galicia, 1 Catalonia and 1 Canaries; Oct–Jan).

Dwarf Bittern *Ixobrychus sturmii* (avetorillo plomizo). Africa. **2** (Canaries: Gran Canaria, Jan 00; Tenerife, Aug 02).

Green-backed Heron *Butorides striata* (garcita verdosa). Americas, Africa, S Asia, New Guinea and Australia. **1**, subsp. *virescens*, American (Canaries: Tenerife, May 06).

Cape Petrel

Cape Verde/
Desertas Petrel

Green-backed Heron

Pelícano común

Great Blue Heron *Ardea herodias* (garza azulada). N America. **1** (Canaries: Tenerife, Dec 98).

Tricolored Heron *Egretta tricolor* (garceta tricolor). Americas. **1** (Canaries: Gran Canaria and Tenerife, Nov 07–May 08).

Western Reef-egret *Egretta gularis* (garceta dimorfa). Africa and S Asia. **46** (15 Andalucía, 11 Levant, 6 Catalonia, 3 Ebro, 3 Meseta, 1 North, 2 Balearics and 5 Canaries; most Apr–Aug. A further 82 accepted as hybrids with *E. garzetta*.

Pink-backed Pelican *Pelecanus rufescens* (pelícano rosado). Africa; some may be escapes. **13** (5 Ebro, 3 Meseta, 2 Catalonia, 1 Andalucía, 1 North and 1 Balearics; year-round).

Great White Pelican *Pelecanus onocrotalus* (pelícano común). SE Europe, Africa and Asia; some may be escapes. **29** (11 Andalucía, 10 Meseta, 3 Ebro, 2 Catalonia, 2 North and Galicia; year round, peak Aug–Oct).

Magnificent Frigatebird *Fregata magnificens* (rabihorcado magnífico). Tropical oceans (nearest colonies in Cape Verdes (almost extinct) and Caribbean). **2** (Andalucía: Torremolinos, Oct 00. North: Santander, Nov 08).

Red-footed Booby *Sula sula* (piquero patirrojo). Tropical oceans (Nearest colonies in Caribbean). **2** (Peninsula: perhaps same individual in Estepona, Andalucía, Aug 10, and Cape Creus and Estartit, Catalonia, Dec 10. Canaries: at sea between Tenerife and Gran Canaria, Sept 12).

Brown Booby *Sula leucogaster* (piquero pardo). Tropical oceans. **7** (4 Galicia, 2 Andalucía and 1 Canaries; most Jul–Sept).

Masked Booby *Sula dactylatra* (piquero enmascarado). Tropical oceans. **3** (Andalucía: Sotogrande, Oct 86 and Torremolinos, Dec 86, perhaps same bird. North: Bilbao, Sept 03).

Double-crested Cormorant *Phalacrocorax auritus* (cormorán orejudo). N America and Cuba. 1 (Canaries: El Hierro, Oct 12).

Egyptian Plover *Pluvianus aegyptius* (pluvial). Africa. **1** (Levant: Alfás del Pi/Cartagena, May–Jun 08; perhaps escaped).

Pacific Golden Plover *Pluvialis fulva* (chorlito dorado siberiano). Siberia. **10** (5 Catalonia, 3 Galicia, 1 Andalucía and 1 Levant; 9 Jul–Dec, peak Aug, and 1 Mar).

American Golden Plover *Pluvialis dominica* (chorlito dorado americano). N America. **56** (25 Canaries, 16 Galicia, 6 North, 5 Andalucía, 3 Catalonia, and 1 Levant; year-round, with main peak in Sept–Oct and secondary peak in May).

Red-footed Booby

Brown Booby

Pacific Golden Plover

Semipalmated Plover

Semipalmated Plover *Charadrius semipalmatus* (chorlitejo semipalmeado). N America. **5** (4 Galicia and 1 Canaries; May–Dec).

Killdeer *Charadrius vociferus* (chorlitejo culirrojo). Americas. **4** (Galicia; Sept–Feb).

Kittlitz's Plover *Charadrius pecuarius* (chorlitejo pecuario). Africa. **1** (Meseta: Villacañas, Mar 01).

Lesser Sandplover *Charadrius mongolus* (chorlitejo mongol chico). Asia. **2** (Meseta: Alcázar de San Juan, Jun 81. Canaries: La Palma, Aug 03).

Greater Sandplover *Charadrius leschenaultii* (chorlitejo mongol grande). Asia. **3** (Catalonia: Ebro delta, Aug 93 and Aug 96. Galicia: Ponteceso, Jul 07).

Sociable Lapwing *Vanellus gregarius* (avefría sociable). Asia and SE Russia. **51** (18 Meseta, 13 Catalonia, 10 Andalucía, 7 Ebro, 1 Levant, 1 Balearics and 1 Canaries; Sept–Apr, most Oct–Mar).

White-tailed Lapwing *Vanellus leucurus* (avefría coliblanca). SO Asia. **2** (Canaries: Tenerife, Nov 78. North: Avilés, Sept 03).

Upland Sandpiper *Bartramia longicauda* (correlimos batitú). N America. **4** (2 North, 1 Galicia and 1 Canaries; Sept–Nov).

Great Knot *Calidris tenuirostris* (correlimos grande). NW Siberia. **1** (Catalonia: Ebro delta, Apr 79).

Broad-billed Sandpiper *Calidris falcinellus* (correlimos falcinelo). N Eurasia. **37** (22 Catalonia, 5 Andalucía, 4 Levant, 4 North, 1 Galicia and 1 Meseta; peaks in Apr–May and Jul–Oct).

Sharp-tailed Sandpiper *Calidris acuminata* (correlimos acuminado). N Siberia. **3** (Levant: Albufera de Valencia, Sept 01 and Sept–Oct 10. North: ría de Villaviciosa, Aug 05).

Stilt Sandpiper *Calidris himantopus* (correlimos zancolín). N America. **1** (Balearics: Mallorca, May 83).

Red-necked Stint *Calidris ruficollis* (correlimos cuellirrojo). E Asia, boreal. **1** (Levant: Albufera de Valencia, Jul 08).

Baird's Sandpiper *Calidris bairdii* (correlimos de Baird). N America and NE Siberia. **17** (8 Galicia, 5 Canaries, 1 Catalonia, 1 Levant, 1 Meseta and 1 Balearics; most Sept–Nov).

Least Sandpiper *Calidris minutilla* (correlimos menudillo). N America. **3** (2 Galicia: Laxe, Oct 87 and Malpica, Oct 99. Catalonia: Ebro delta, Aug 96).

White-rumped Sandpiper *Calidris fuscicollis* (correlimos culiblanco). N America. **67** (41 Canaries, 10 Galicia, 4 Andalucía, 3 Catalonia, 3 Levant, 3 North, 1 Ebro, 1 Meseta and 1 Balearics; most Aug–Nov).

Killdeer

Greater Sandplover

Sociable Lapwing

Broad-billed Sandpiper

Buff-breasted Sandpiper *Calidris subruficollis* (correlimos canelo). N America. **96** (25 Galicia, 18 Catalonia, 16 Canaries, 14 Levant, 9 North, 7 Meseta, 6 Andalucía and 1 Balearics; most Sept–Oct).

Pectoral Sandpiper *Calidris melanotos* (correlimos pectoral). N America and Siberia. **346** (287 Peninsula, widely spread, with most in Galicia, North, Catalonia and Levant, and 59 Canaries; esp. Jul–Oct, peaking Sept, and a few in Apr–May). (*)

Semipalmated Sandpiper *Calidris pusilla* (correlimos semipalmeado). N America. **16** (8 Galicia, 4 Canaries, 3 North and 1 Catalonia; 14 Jul–Nov, peaking Sept, and 2 May–Jun).

Western Sandpiper *Calidris mauri* (correlimos de Alaska). Alaska and NE Siberia. **3** (Galicia: Laxe, Sept 79 and Sept 99; Baldaio, Sept 92).

Short-billed Dowitcher *Limnodromus griseus* (agujeta gris). N America. **1** (Levant: salinas de San Pedro del Pinatar, Mar 13).

Long-billed Dowitcher *Limnodromus scolopaceus* (agujeta escolopácea). NE Siberia and extreme NW N America. **21** (17 in Peninsula, including 7 Galicia and 4 Catalonia; 1 Balearics and 3 Canaries; most Aug–Dec, peaking Oct, some Feb–May).

Great Snipe *Gallinago media* (agachadiza real). N Europe and NW Asia. **33** (17 Catalonia, 10 Balearics, 2 Galicia, 2 Meseta, 1 Ebro and 1 Levant; Feb–May, peaks in Apr, and Aug–Dec, esp. Sept–Oct). (*)

Wilson's Snipe *Gallinago delicata* (agachadiza de Wilson). N America. **1** (Canaries: Tenerife, Oct–Nov 13).

Wilson's Phalarope *Steganopus tricolor* (falaropo tricolor). N America. **26** (20 Península, evenly distributed, 1 Balearics and 5 Canaries; year-round, peaking Apr–May and Aug–Oct).

Terek Sandpiper *Xenus cinereus* (andarríos del Terek). N Eurasia. **58** (19 Andalucía, 19 Catalonia, 9 Levant, 4 North, 2 Balearics, 2 Ebro, 1 Galicia, 1 Meseta and 1 Canaries; year-round, peaking May and Aug–Sept).

Spotted Sandpiper *Actitis macularius* (andarríos maculado). N America. **35** (17 Canaries, 7 Galicia, 6 North, 3 Catalonia, 1 Andalucía and 1 Meseta; Aug–May, most Aug–Dec).

Solitary Sandpiper *Tringa solitaria* (andarríos solitario). N America. **1** (North: Cazanes, Asturias, Oct 05).

Lesser Yellowlegs *Tringa flavipes* (archibebe patigualdo chico). N America. **75** (22 Galicia, 10 Catalonia, 7 Levant, 7 Meseta, 6 Andalucía, 5 North, 1 Ebro, 2 Balearics and 15 Canaries; year-round, peaking Sept–Nov and Apr–May).

Buff-breasted Sandpiper

Western Sandpiper

Terek Sandpiper

Lesser Yellowlegs

Greater Yellowlegs *Tringa melanoleuca* (archibebe patigualdo grande). N America. **9** (5 Galicia, 2 Andalucía, 1 Balearics and 1 Canaries; year-round).

Cream-coloured Courser *Cursorius cursor* (corredor sahariano). Africa (resident in Canaries, p. 96, and has bred sporadically in the Peninsula). Otherwise **23** Peninsula (15 Andalucía, 3 Meseta, 2 Catalonia and 2 Galicia; 12 Mar–Jun, peak Apr, and 11 Aug–Dec, peak Sept).

Black-winged Pratincole *Glareola nordmanni* (canastera alinegra). Asia and extreme E Europe. **3** (Balearics: Mallorca, Apr 91 and May 05. Levant: Albufera de Valencia, Aug–Sept 07).

Ross's Gull *Rhodostethia rosea* (gaviota rosada). NE Siberia. **3** (North: Hondarribia and Zumaia, Apr 94. Meseta: Alcázar de San Juan, Feb–Mar 09).

Bonaparte's Gull *Larus philadelphia* (gaviota de Bonaparte). N America. **33** (15 Galicia, 12 North, 2 Andalucía, 2 Meseta, 1 Catalonia and 1 Canaries; esp. Jan–Mar).

Grey-headed Gull *Larus cirrocephalus* (gaviota cabecigrís). Africa and S America. **2** (Andalucía: Marismas del Guadalquivir, Jun 71. Canaries: Las Palmas de Gran Canaria, Feb 05).

Franklin's Gull *Larus pipixcan* (gaviota pipizcan). N America. **18** (5 Andalucía, 4 Catalonia, 3 North, 2 Meseta, 1 Galicia, 1 Levant, and 2 Canaries; year-round).

Laughing Gull *Larus atricilla* (gaviota guanaguanare). N America. **50** (15 Galicia, 9 Catalonia, 9 North, 7 Andalucía, 1 Levant and 2 Canaries; year-round, peak Nov–Jan).

Ring-billed Gull *Larus delawarensis* (gaviota de Delaware). N America. **192** records up to 2005 (110 North, 49 Galicia, 8 Andalucía, 6 Catalonia, 3 Levant and 16 Canaries; year-round, most Dec–Mar). (*)

Kelp Gull *Larus dominicanus* (gaviota cocinera). Sudamérica, S Africa, Australia and New Zealand. **1** (Canaries: Tenerife, Apr 01).

Caspian Gull *Larus cachinnans* (gaviota del Caspio). SE Europe and SW Asia. **78** since 2005 (26 Catalonia, 19 Galicia, 17 North, 11 Meseta, 1 Levant and 4 Balearics; most Dec–Mar). (*)

American Herring Gull *Larus smithsonianus* (gaviota argéntea americana). N America. **5** (2 Galicia, 2 North and 1 Meseta; Dec–Mar).

Thayer's Gull *Larus thayeri* (gaviota esquimal). N Canada, boreal. **1** (Galicia: Cervo, Lugo, same bird returning winters 2008–2011).

Glaucous-winged Gull *Larus glaucescens* (gaviota de Bering). NE Asia and NW N America, boreal. **1** (Canaries: El Hierro, Feb 92).

Bonaparte's Gull

Laughing Gull

Ring-billed Gull

Iceland Gull

Iceland Gull *Larus glaucoides* (gaviota groenlandesa). NE Canada and Greenland. **107** (nearly all in Galicia and North, but 3 in Canaries; year-round, most Dec–Apr). (*)

Glaucous Gull *Larus hyperboreus* (gavión hiperbóreo). Eurasia and N America, boreal. **170** (great majority in North and Galicia, but 6 in Canaries; year-round, peak Dec–Mar). (*)

Sooty Tern *Onychoprion fuscatus* (charrán sombrío). Tropical seas. **4** (2 Galicia, 1 Catalonia, 1 Canaries; Jun–Sept).

Bridled Tern *Onychoprion anaethetus* (charrán embridado). Tropical seas. **1** (Catalonia: Ebro delta, Jun 08).

Forster's Tern *Sterna forsteri* (charrán de Forster). N America. **6** (4 Andalucía and 2 Galicia; Aug–Jan).

Elegant Tern *Thalasseus elegans* (charrán elegante). N America, Pacific coast. **1** (Catalonia: Delta del Llobregat, Apr 93).

Royal Tern *Thalasseus maximus* (charrán real). Americas and W Africa. **24** (20 Andalucía, 4 Catalonia; most Jul–Oct). P. 106.

Long-tailed Skua *Stercorarius longicaudus* (págalo rabero). Eurasia and N America, boreal. **58** (20 North, 14 Galicia, 11 Mediterranean coast, 6 peninsular interior, 3 Andalucía, 1 Balearics and 3 Canaries; Apr–Nov, most Aug–Oct). No longer treated as a rarity; regular migrant off NW coast in autumn.

South Polar Skua *Catharacta maccormicki* (págalo polar). Antarctic. **2** (Canaries: Banco de la Concepción, Sept 11. Galicia: off Vivero, Oct 13).

Black Guillemot *Cepphus grylle* (arao aliblanco). N Atlántico and Arctic. **4** (2 Galicia and 2 North; Aug–Dec).

Marsh Owl *Asio capensis* (búho moro). Africa, inc. Morocco. **1** (Andalucía: Jerez de la Frontera, Dec 98).

Swallow-tailed Kite *Elanoides forficatus* (elanio tijereta). Americas. **1** (Canaries: Fuerteventura, Mar 93).

Bateleur Eagle *Terathopius ecaudatus* (águila volatinera). Africa. **1** (Andalucía: Algeciras, Apr 12).

White-backed Vulture *Gyps africanus* (buitre dorsiblanco africano). Africa. **3** (Andalucía: Strait region, Sept 08, Jun 09 and Sept 11).

Rüppell's Vulture *Gyps rueppellii* (buitre moteado). Africa. **77** (51Andalucía, 21 Meseta, 2 Ebro, 2 Levant and 1 Ceuta; year-round, most Sept–Oct in Andalucía). (*)

Lesser Spotted Eagle *Clanga pomarina* (águila pomerana). E Europe. **8** (5 Andalucía, 2 Balearics and 1 Catalonia; esp. Sept–Nov).

Glaucous Gull

Sooty Tern

Royal Tern

Rüppell's Vulture

Greater Spotted Eagle *Aquila clanga* (águila moteada). N Europe and Siberia. **23** (10 Levant, 8 Andalucía, 4 Catalonia and 1 Galicia; Sept–May, most Oct–Mar).

Steppe Eagle *Aquila nipalensis* (águila esteparia). SE Europe, W and C Asia. **1** (Andalucía: Tarifa, May 08).

Eastern Imperial Eagle *Aquila heliaca* (águila imperial oriental). E Europe and Asia. **1** (Levant: ringed bird found dead in sierra de Escalona, Alicante, Mar 10).

Pallid Harrier *Circus macrourus* (aguilucho papialbo). E Europe and Asia. **67** (14 Catalonia, 12 Levant, 11 Andalucía, 11 Ebro, 9 Balearics, 4 Meseta, 2 North and 4 Canaries; Aug–May, peaking Sept and esp. Mar–Apr). (*)

Pallid Harrier

White-tailed Eagle *Haliaeetus albicilla* (pigargo europeo). N and E Europe and Asia. **9** (4 Ebro, 2 North, 1 Catalonia, 1 Ebro and 1 Meseta; Dec–Mar).

Rough-legged Buzzard *Buteo lagopus* (busardo calzado). Eurasia and N America, boreal. **7** (Catalonia; Oct–Mar).

Long-legged Buzzard *Buteo rufinus* (busardo moro). N Africa, SE Europe and Asia. **38** (24 Andalucía, 6 Levant, 2 Meseta, 1 Catalonia, 1 Ebro, 1 Galicia and 3 Canaries; year-round, peak Sept). Also observed in Ceuta, where not a vagrant. P. 134.

Rough-legged Buzzard

Blue-cheeked Bee-eater *Merops persicus* (abejaruco persa). Africa and SW Asia. **4** (Andalucía, Balearics, Catalonia and Ebro; Apr, Jun and Sept).

Belted Kingfisher *Megaceryle alcyon* (martín gigante norteamericano). N America. **1** (Levant: Cartagena, Nov 09–Mar 10).

Lanner Falcon *Falco biarmicus* (halcón borní). SE Europe, SW Asia and Africa; some possibly escapes. **14** (7 Andalucía, 3 Catalonia, 3 Meseta and 1 Levant; 10 Aug–Oct).

Lanner Falcon

Saker Falcon *Falco cherrug* (halcón sacre). E Europe and Asia. **2** (Andalucía: Tarifa/Algeciras, Apr 02. Meseta: Becilla de Valderaduey/ Villalón de Campos, Aug 09).

Gyrfalcon *Falco rusticolus* (halcón gerifalte). Eurasia and N America, boreal. **1** (North: Tapia de Casariego, Asturias, Dec 09–Mar 10).

Fork-tailed Flycatcher *Tyrannus savana* (tijereta sabanera). C and S America. **1** (Andalucía: El Rocío, Oct 02).

Red-eyed Vireo *Vireo olivaceus* (vireo chiví). N America. **3** (Catalonia: Tarragona, Oct 95; Tiana, Oct 00. Levant: Elche, Oct 95).

Black-crowned Tchagra *Tchagra senegalus* (chagra del Senegal). Africa, including Morocco, and Arabia. **1** (Andalucía: Tarifa, Jul

Isabelline Shrike

95) and various records in Ceuta, where not a vagrant.

Isabelline Shrike *Lanius isabellinus* (alcaudón isabel). Asia. **5** (2 Andalucía, 1 Balearics, 1 Galicia and 1 Levant; Sept–Dec and Feb).

Masked Shrike *Lanius nubicus* (alcaudón núbico). SE Europe and SW Asia. **2** (Balearics: Mallorca, Apr 91, and Cabrera, Sept 03).

Spotted Nutcracker *Nucifraga caryocatactes* (cascanueces común). Eurasia. **1** (Catalonia: El Serrat, Feb 85).

Hooded Crow *Corvus cornix* (corneja cenicienta). Europe, W Asia and Egypt. **13** (5 Catalonia, 4 Galicia, 3 North and 1 Ceuta; 10 Mar–Jun and 3 Nov).

Greater Hoopoe-lark *Alaemon alaudipes* (alondra ibis). N Africa and SW Asia. **5** (1 Chafarinas Is, Aug, and 4 Canaries, Jan–Mar).

Bar-tailed Lark *Ammomanes cinctura* (terrera colinegra). N Africa and SW Asia. **5** (1 Balearics, 1 Catalonia, 1 Levant and 2 Canaries; Feb–May).

Horned Lark *Eremophila alpestris* (alondra cornuda). N and E Europe, NW Africa, Asia and Americas. **4** (3 North, 1 Ebro; Oct–Mar).

Booted Warbler *Iduna caligata* (zarcero escita). NE Europe and W Asia. **3** (Balearics: Dragonera, Sept 98. North: Colunga, Asturias, Sept 04. Galicia: Ferrol, Oct 09).

Blyth's Reed-warbler *Acrocephalus dumetorum* (carricero de Blyth). NE Europe and W Asia. **2** (Andalucía: Doñana, same bird in Sept 01 and Aug 02. Balearics: Cabrera, Oct 07).

Marsh Warbler *Acrocephalus palustris* (carricero políglota). Europe and W Asia. **4** (Catalonia, Levant, Balearics; Apr–May and Aug–Sept). P. 170.

Paddyfield Warbler *Acrocephalus agricola* (carricero agrícola). Asia. **13** (8 Catalonia, 2 Andalucía, 1 Levant and 1 Meseta; Aug–Apr).

River Warbler *Locustella fluviatilis* (buscarla fluvial). C and E Europe and W Asia. **1** (Galicia: Miño estuary, Sept 03).

Common Bulbul *Pycnonotus barbatus* (bulbul naranjero). Africa. **5** (Andalucía, where has bred; year-round; escaped birds a possibility).

Yellow-browed Warbler *Abrornis inornata* (mosquitero bilistado). Asia. **156** (30 Andalucía, 23 Balearics, 17 Catalonia, 17 Levant, 12 North, 5 Meseta, 4 Galicia, 1 Ceuta and 44 Canaries; esp. Oct–Nov in Peninsula and Balearics and Jan–Mar in Canaries). (*)

Hume's Leaf-warbler *Abrornis humei* (mosquitero de Hume). Asia central. **2** (Meseta: Cabañas del Castillo, Cáceres, Feb 06. Galicia: Ribadeo, Jan–Feb 12).

Greater Hoopoe-lark

Marsh Warbler

Paddyfield Warbler

Yellow-browed Warbler

Pallas's Leaf-warbler *Abrornis proregulus* (mosquitero de Pallas). Asia. **14** (4 Levant, 2 Andalucía, 2 Balearics, 2 Catalonia, 2 Meseta, 1 North and 1 Chafarinas Is; 8 Feb–Apr, 6 Oct–Dec).

Dusky Warbler *Phylloscopus fuscatus* (mosquitero sombrío). Asia. **6** (3 Catalonia, 2 Galicia and 1 Balearics; 1 May and 5 Oct–Nov).

Radde's Warbler *Phylloscopus schwarzi* (mosquitero de Schwarz). Asia. **7** (3 Balearics, 2 North, 1 Andalucía and 1 Levant; Oct–Nov).

Greenish Warbler *Seicercus trochiloides* (mosquitero verdoso). Eurasia. **3** (Levant: Columbretes Is, Sept 88. Catalonia: Llobregat delta, Sept 04. Balearics: Menorca, Sept 13).

Arctic Warbler *Seicercus borealis* (mosquitero boreal). N Eurasia. **1** (Andalucía: laguna de Medina, Sept 08).

Desert Warbler *Sylvia nana* (curruca sahariana). N Africa to C Asia. **8** (*S. n. deserti*, African: 6 Canaries and 1 Levant; 4 Feb–Apr and 2 Oct–Nov. *S. n. nana*, Asian: 1 in Ajo, Cantabria, Nov 13).

Barred Warbler *Sylvia nisoria* (curruca gavilana). Eurasia. **4** (2 Balearics, 1 Levant and 1 North; Sept–Oct).

Tristram's Warbler *Sylvia deserticola* (curruca del Atlas). NW Africa. **1** (Canaries: Fuerteventura, Oct 95).

Rüppell's Warbler *Sylvia ruppeli* (curruca de Rüppell). SE Europe and SW Asia. **1** (Balearics: Cabrera, Apr 10).

Marmora's Warbler *Sylvia sarda* (curruca sarda). Corsica and Sardinia. **4** (2 Catalonia and 2 Balearics; Mar–May).

Rose-coloured Starling *Pastor roseus* (estornino rosado). SE Europe and SW Asia. **37** (9 Catalonia, 8 North, 6 Balearics, 6 Levant, 3 Andalucía, 2 Ebro, 1 Galicia, 1 Meseta and 1 Canaries; year-round, peaking Sept–Nov and Apr–May).

Grey Catbird *Dumetella carolinensis* (pájaro gato gris). N America. **1** (Canaries: Tenerife, Nov 99).

American Robin *Turdus migratorius* (zorzal robín). N America. **1** (North: Laukiz, Vizcaya, Dec 99).

Naumann's Thrush *Turdus naumanni* (zorzal de Naumann). Asia. **1** (Catalonia: Ebro delta, Oct 05).

Siberian Blue Robin *Larvivora cyane* (ruiseñor azul). Asia. **1** (Catalonia: Ebro delta, Oct 00).

Orange-flanked Bush-robin *Tarsiger cyanurus* (ruiseñor coliazul). Siberia. **6** (2 Catalonia, 2 Levant, 1 Andalucía and 1 Meseta; Oct–Mar).

Red-breasted Flycatcher *Ficedula parva* (pa-

Pallas's Leaf-warbler

Dusky Warbler

Desert Warbler

Rose-coloured Starling

pamoscas papirrojo). C and E Europe, Asia. **41** (12 Levant, 8 Balearics, 4 Andalucía, 3 Catalonia, 3 Meseta, 2 Galicia, 2 North and 7 Canaries; most Sept–Nov).

Semi-collared Flycatcher *Ficedula semitorquata* (papamoscas semiacollarado). SE Europe and SW Asia. **1** (Balearics: Cabrera, Apr 11).

Collared Flycatcher *Ficedula albicollis* (papamoscas acollarado). C and E Europe, SW Asia. **60** (34 Balearics, 18 Catalonia, 5 Levant, 3 Andalucía, 3 Catalonia, 3 Meseta, 2 Galicia, 2 North and 1 Canaries; most Apr–May).

Moussier's Redstart *Phoenicurus moussieri* (colirrojo diademado). NW Africa. **5** (2 Andalucía, 2 Balearics and 1 Catalonia; 1 Jan, 3 Apr and 1 Oct).

Isabelline Wheatear *Oenanthe isabellina* (collalba isabel). SE Europe and Asia. **7** (5 Galicia, 1 Balearics and 1 Canaries; Oct–Nov).

Desert Wheatear *Oenanthe deserti* (collalba desértica). N Africa and SW Asia. **25** (12 Canaries, 3 Andalucía, 3 Galicia, 2 Catalonia, 2 North, 1 Ebro, 1 Levant and 1 Balearics; Aug–Apr).

White-crowned Wheatear *Oenanthe leucopyga* (collalba yebélica). N Africa and Arabia. **1** (Canaries: La Palma, Jan 05).

Bohemian Waxwing *Bombycilla garrulus* (ampelis europeo). N Eurasia and northern N America. **14** (7 North, 2 Catalonia, 2 Galicia, 2 Meseta and 1 Ebro; Nov–Apr).

Buff-bellied Pipit *Anthus rubescens* (bisbita norteamericano). N America. **1** (Galicia: A Coruña, Sept 11).

Olive-backed Pipit *Anthus hodgsoni* (bisbita de Hodgson). Asia. **10** (6 Balearics, 2 Andalucía, 1 Galicia and 3 Canaries; Oct–Apr, 7 in Oct).

Richard's Pipit *Anthus richardi* (bisbita de Richard). Asia. **84** up to 2005 (31 Galicia, 25 North, 8 Catalonia, 7 Balearics, 6 Andalucía, 3 Levant and 4 Canaries; Sept–Jun, esp. Oct–Apr). P. 206. (*)

Blyth's Pipit *Anthus godlewskii* (bisbita estepario). Asia. **1** (North: Tapia, Nov–Dec 05).

Citrine Wagtail *Motacilla citreola* (lavandera cetrina). E Europe and Asia. **73** (35 Catalonia, 18 Levant, 5 Balearics, 3 Andalucía, 1 Galicia, 1 Meseta and 10 Canaries; Aug–May, peaking Sept and Apr).

Common Rosefinch *Erythrina erythrina* (camachuelo carminoso). Eurasia. **36** (17 Balearics, 6 Andalucía, 3 Catalonia, 3 North, 2 Levant, 2 Meseta, 1 Ebro, 1 Ceuta and 1 Canaries; 33 Sept–Nov, most Sept, and 3 Apr–Jun).

Twite *Carduelis flavirostris* (pardillo piquigualdo). N Europe and Asia. **1** (North: Gijón, Oct 93).

Collared Flycatcher

Moussier's Redstart

Desert Wheatear

Common Rosefinch

Common Redpoll *Carduelis flammea* (pardillo sizerín). Eurasia. **7** (2 Ebro, 1 Andalucía, 1 Galicia, 1 Levant, 1 North and 1 Balearics; Nov–Jan).

Lapland Bunting *Calcarius lapponicus* (escribano lapón). Eurasia and N America, boreal. **50** (24 North, 23 Galicia, 1 Catalonia, 1 Levant and 1 Balearics; Sept–Jan, peak Oct).

House Bunting *Fringillaria sahari* (escribano sahariano). N Africa. **2** (Andalucía: Algeciras, Jun 87; Nerja, Oct 09).

Red-headed Bunting *Granivora bruniceps* (escribano carirrojo). C Asia. **4** (3 Catalonia and 1 Andalucía; Oct–Nov).

Black-headed Bunting *Granivora melanocephala* (escribano cabecinegro). SE Europe and SW Asia. **9** (4 Catalonia, 3 Balearics and 2 Andalucía; 7 Apr–Jun and 2 Oct).

Cretzschmar's Bunting *Emberiza caesia* (escribano ceniciento). SE Europe and SW Asia. **1** (Balearics: Cabrera, May 11).

Pine Bunting *Emberiza leucocephalos* (escribano cabeciblanco). Asia. **5** (2 Catalonia, 2 Levant and 1 Meseta; Oct–Mar).

Rustic Bunting *Schoeniclus rusticus* (escribano rústico). N Europe and N Asia. **7** (3 Andalucía, 2 Catalonia, 1 Levant and 1 Meseta; 5 Oct–Dec and 2 Feb–Mar).

Little Bunting *Schoeniclus pusillus* (escribano pigmeo). NE Europe and Siberia. **58** (12 Catalonia, 11 Balearics, 9 Meseta, 6 North, 5 Galicia, 5 Levant, 4 Andalucía, 4 Ebro and 2 Canaries; Sept–Apr, most Oct–Dec).

Yellow-breasted Bunting *Schoeniclus aureolus* (escribano aureolado). Eurasia, boreal. **2** (Ebro: Belchite, Nov 88. Balearics: Cabrera, Oct 95).

Louisiana Waterthrush *Parkesia motacilla* (reinita charquera de Luisiana). N America. **1** (Canaries: La Palma, Nov 91).

Black-and-white Warbler *Mniotilta varia* (reinita trepadora). N America. **1** (Canaries: Fuerteventura, Dec 09).

Yellow-rumped Warbler *Setophaga coronata* (reinita coronada). N and C America. **2** (Galicia: Laxe, Sept 82. Canaries: Gran Canaria, Feb–Mar 84).

Bobolink *Dolichonyx oryzivorus* (charlatán). N America. **1** (Canaries: Lanzarote, Oct–Dec 10).

Common Redpoll

Lapland Bunting

Black-headed Bunting

Little Bunting

AVIFAUNA

Adrover, J. & others (2010). *Atles dels aucells nidificants de Mallorca i Cabrera (2003–2007).* GOB. Palma de Mallorca.
Aierbe, T., Olano, M. & Vázquez, J. (2001). *Atlas de las aves nidificantes de Gipuzkoa.* Sociedad de Ciencias Aranzadi. Donostia-San Sebastián.
Álvarez, J., & others (1985). *Atlas de los vertebrados continentales de Álava, Vizcaya y Guipúzcoa.* Departamento de Política Territorial y Transportes, Gobierno Vasco. Vitoria.
Álvarez-Balbuena, F. & others (2000). *Aves raras y escasas en Asturias.* Coordinadora Ornitolóxica d'Asturies. Avilés.
Aragüés, A. & Lucientes, J. (1980). *Fauna de Aragón: las aves.* Guara Editorial. Zaragoza.
Arcos, J. M. & others (2009). *Áreas importantes para la conservación de las aves marinas en España.* SEO/BirdLife. Madrid.
Associació per a la Defensa de la Natura (2002). *Atles dels ocells nidificants d'Andorra.* Associació per a la Defensa de la Natura. Andorra la Vella.
Aymerich, J. & others (1991). *Els ocells d'Osona.* Lynx Edicions. Barcelona.
Bahillo, M. & Alonso, I. (1998). *Fauna de Cantabria. Aves acuáticas y marinas.* Cantabria Tradicional. Santander.
Bernis, F. (1966–1971). *Aves Migradoras Ibéricas*, vols. 1–8. Sociedad Española de Ornitología. Madrid.
Bernis, F. (1980). *La migración de las aves en el Estrecho de Gibraltar*, volumen I: Aves Planeadoras. Cátedra de Zoología de Vertebrados, Universidad Complutense. Madrid.
Bueno, R. (2000). *Las aves de la sierra de Béjar.* Ed. Raúl Bueno. Béjar.
Calvet, J. & others (eds.) (2004). *Els ocells de la Plana de Lleida.* Pagès editors. Lleida.
Campos, B. (ed.) (2001). *Atlas de las aves nidificantes del Alto Vinalopó (Alicante).* Asociación Villenense de Amigos de la Naturaleza. Villena (Alicante).
Carnero, I. & Peris, S. (1988). *Atlas ornitológico de la provincia de Salamanca.* Diputación de Salamanca. Salamanca.
Carrascal, L. M. & Palomino, D. (2008). *Las aves comunes reproductoras en España. Población en 2004–2006.* SEO/Birdlife. Madrid.
Castro, H. (1993). *Ecología y dinámica anual de las poblaciones de aves en las salinas de Cabo de Gata (Almería).* Instituto de Estudios Almerienses. Almería.
Catry, P. & others (2010). *Aves de Portugal. Ornitologia do território continental.* Assírio & Alvim. Lisbon.
Ceballos, J. J. & Guimerá, V. M. (1992). *Guía de las aves de Jerez y de la provincia de Cádiz. Atlas ornitológico de las especies nidificantes.* Biblioteca de Urbanismo y Cultura. Jerez de la Frontera.
Ceña, A., Ceña, J. C. & Moya, I. (eds.) (1996). *Fauna de La Rioja (Vertebrados).* Caja de Ahorros de La Rioja. Logroño.
Chiclana, F., Lama, J. A. & Salcedo, J. (2002). *Aves de la provincia de Sevilla.* Diputación Provincial de Sevilla y SEO-Sevilla. Sevilla.
Clarke, T. (2006). *A field guide to the birds of the Atlantic Islands: Canary Islands, Madeira, Azores, Cape Verde.* Christopher Helm. London.
Clavell, J. (2002). *Catàleg dels ocells dels Països Catalans.* Lynx Edicions. Barcelona.
Cordero-Tapia, P. J. (1983). *Las aves del Maresme. Catálogo, status y fenología.* Edicions Universitat de Barcelona. Barcelona.
Cramp, S. (ed.) (1977–1994). *Birds of the Western Palearctic: handbook of the birds of Europe, the Middle East and North Africa.* Vols. 1–9. Oxford University Press. Oxford.
De Juana, E. (1980). *Atlas ornitológico de La Rioja.* Instituto de Estudios Riojanos. Logroño.
De Juana, E. (2006). *Aves raras de España.* Lynx Edicions. Barcelona.
De Juana, E. & Garcia, E. (2015). *The birds of the Iberian Peninsula.* Christopher Helm. London.

Del Hoyo, J. & others (eds). (1992–2011). *Handbook of the Birds of the World*. 16 volumes. Lynx Edicions. Barcelona.
Del Moral, J. C. & others (eds.) (2002). *Atlas de las aves invernantes de Madrid 1999-2001.* SEO-Monticola and Comunidad de Madrid. Madrid.
Del Moral, J. C. & others (eds.) (2012). *Atlas de las aves en invierno en España 2007–2010.* Ministerio de Agricultura, Alimentación y Medio Ambiente, and SEO/BirdLife. Madrid.
De Moral, J.C. & others (2014). *Programas de seguimiento de avifauna de SEO/BirdLife 2013.* SEO/BirdLife. Madrid.
Díaz, M., Asensio, B. & Tellería, J. L. (1996). *Aves ibéricas I. No paseriformes.* J. M. Reyero Editor. Madrid.
Dies, B. & others (1999). *Las aves de l'Albufera de Valencia.* Vaersa. Valencia.
Dubois, P. J. & others (2008). *Nouvel inventaire des oiseaux de France*. Delachaux et Niestlé. Paris.
Elósegui, J. (1985). *Navarra. Atlas de aves nidificantes.* Caja de Ahorros de Navarra. Pamplona.
Emmerson, K. & others (1994). *Catálogo y bibliografía de la avifauna canaria.* Museo de Ciencias Naturales, Cabildo de Tenerife. Santa Cruz de Tenerife.
Equipa Atlas (2008). *Atlas das aves nidificantes em Portugal (1999–2005).* Assirio & Alvim. Lisbon.
Escandell, A. (1997). *Atles dels ocells nidificants de Menorca.* GOB Menorca. Maó.
Estrada, J. & others (eds.) (2004). *Atles dels ocells nidificants de Catalunya 1999-2002.* Institut Català d'Ornitologia/Lynx Edicions. Barcelona.
Ferrer, X., Martínez-Vilalta, A. & Muntaner, J. (1986). *Història Natural dels Països Catalans*. Vol. 12, Ocells. Enciclopèdia Catalana S. A. Barcelona.
Finlayson, C. (1992). *Birds of the Strait of Gibraltar.* T. & A.D. Poyser. London.
Finlayson, J. C. & Cortés, J. E. (1987). *The birds of the Strait of Gibraltar. Its waters and northern shore.* Alectoris, 6. The Gibraltar Ornithological and Natural History Society. Gibraltar.
Gacio, H. & Sayago, J. M. (2011). *Guía de las aves del Paraje Natural Marismas del Odiel (Huelva).* Algakon. Sevilla.
Gainzarain, J. A. (2006). *Atlas de las aves invernantes en Álava (2002–2005).* Diputación Foral de Álava. Vitoria.
Galarza, A. (2001). *Las aves del Parque Natural de Gorbeia.* Diputación Foral de Álava. Vitoria.
Galarza, A. & Domínguez, A. (1989). *Urdaibai. Avifauna de la Ría de Gernika.* Diputación Foral de Bizkaia, Departamento de Agricultura. Bilbao.
García, E. & others (2014). *Atlas de las aves nidificantes de Asturies (1990–2010).* Coordinadora Ornitolóxica d'Asturies e INDUROT. Llanera (Asturias).
García, J., Ramos, L. A. & Vázquez, X. (2008). *Atlas de las aves nidificantes de la provincia de León.* Edilesa. León.
García, L., Calderón, J. & Castroviejo, J. (1989). *Las Aves de Doñana y su entorno, con especial referencia a la invernada 1985/86.* Cooperativa Marismas del Rocío. El Rocío (Huelva).
García, L. & others (2000). *Prontuario de las aves de Doñana. Anuario Ornitológico de Doñana Diciembre 2000.* Estación Biológica de Doñana y Ayuntamiento de Almonte. Almonte.
Garrido Guil, H. (1996). *Aves de las Marismas del Odiel y su entorno (con especial referencia a los no-paseriformes).* Editorial Rueda. Alcorcón (Madrid).
Garrido, M. & Alba, E. (1997). *Las aves de la provincia de Málaga.* Diputación Provincial de Málaga. Málaga.
Garrido, M., Alba, E. & González, J. M. (2002). *Las aves acuáticas y marinas de Málaga y su provincia*. Centro de Ediciones de la Diputación de Málaga. Málaga.
Garzón, J. & Henares, I. (2012). *Las Aves de Sierra Nevada.* Consejería de Medio Ambiente de la Junta de Andalucía. Granada.

González, R. & Pérez-Aranda, D. (2011). *Las aves acuáticas en España, 1980–2009.* SEO/Birdlife. Madrid.
González-Cachinero, J. M. & others (2006). *Aves acuáticas de la provincia de Granada.* CajaGRANADA, Obra Social. Granada.
Gutiérrez, R., de Juana, E. & Lorenzo, J. A. (2012). *Lista de las aves de España. Edición de 2012.* Online version 1.0. SEO/BirdLife. Madrid.
Gutiérrez, R., Esteban, P. & Santaeufemia, F. X. (1995). *Els ocells del delta del Llobregat.* Lynx. Barcelona.
Hagemeijer, W. J. M. & Blair, M. J. (eds.) (1997). *The EBCC atlas of European breeding birds: their distribution and abundance.* T. & A.D. Poyser. London.
Herrando, S. & others (eds.) (2011). *Atles dels ocells de Catalunya a l'hivern 2006–2009.* Institut Català d'Ornitologia y Lynx Edicions. Barcelona.
Hidalgo, J. & del Villar, J. (2004). *Urdaibai: uretako hegaztien gidaliburua/guía de aves acuáticas*. Departamento de Ordenación del Territorio y Medio Ambiente, Gobierno Vasco. Vitoria.
Infante, O., Fuente, U. & Atienza, J.C. (2011). *Las Áreas Importantes para la Conservación de las Aves en España.* SEO/BirdLife. Madrid.
Jiménez, J. & Navarrete, J. (2001). *Estatus y fenología de las aves de Ceuta*. Instituto de Estudios Ceutíes. Ceuta.
Jiménez García-Herrera, J. & others (1992). *Las aves del Parque Natural de las Tablas de Daimiel y otros humedales manchegos.* Lynx Edicions. Barcelona.
Jubete, F. (1997). *Atlas de las aves nidificantes de la provincia de Palencia.* Asociación de Naturalistas Palentinos. Palencia.
Jubete, F. & Martín, C. M. (2009). *Aves de las lagunas de La Nava, Boada y Pedraza y su entorno.* Obra Social de Caja España. Palencia.
Llandres, C. & Urdiales, C. (1990). *Las aves de Doñana.* Lynx Edicions. Barcelona.
Madroño, A., González, C. & Atienza, J. C. (eds.) (2004). *Libro rojo de las aves de España*. Dirección General para la Biodiversidad y SEO/BirdLife. Madrid.
Manrique, J. (1993). *Las aves de Almería.* Instituto de Estudios Almerienses. Almería.
Martí, R. & others (coords.) (1994). *Atlas de las aves nidificantes en Madrid.* Sociedad Española de Ornitología y Agencia de Medio Ambiente de la Comunidad de Madrid. Madrid.
Martí, R. & del Moral, J. C. (eds.) (2003). *Atlas de las aves reproductoras de España*. Dirección General de Conservación de la Naturaleza y Sociedad Española de Ornitología. Madrid.
Martín, A. (1987). *Atlas de las aves nidificantes en la isla de Tenerife.* Instituto de Estudios Canarios. Santa Cruz de Tenerife.
Martín, A. & Lorenzo, J. A. (2001). *Aves del Archipiélago Canario*. Francisco Lemus, Editor. La Laguna.
Martín, L. J. & Sierra, G. (1999). *Guía de las aves de La Moraña y Tierra de Arévalo*. ASODEMA. Ávila.
Martínez, R. & others (1996). *Atlas de las aves del norte de Murcia (Jumilla-Yecla), 1989-1993.* Jumilla-Yecla (Murcia).
Martínez Mariño, J. M. (coord.) (2009). *Anuario histórico das aves de Galicia.* Sociedade Galega de Ornitoloxía. Santiago de Compostela.
Martínez-Vilalta, A. & Motis, A. (1989). *Els ocells del Delta de l'Ebre.* Lynx Edicions. Barcelona.
Muntaner, J. & Congost, J. (1984). *Avifauna de Menorca.* Treballs del Museu de Zoologia, Ajuntamente de Barcelona. Barcelona.
Noval, A. (2001). *Guía de las aves de* Asturias, 3ª edición. Alfredo Noval editor. Gijón.
Parejo, E. L. & Sáez, O. (1995). *Estudio ornitológico del Campo de Gibraltar y Ceuta*. Instituto de Estudios Campogibraltareños. Algeciras.
Paterson, A. M. (1997). *Las aves marinas de España y Portugal.* Lynx Edicions. Barcelona.
Pedrocchi-Renault, C. (1987). *Fauna ornítica del Alto Aragón Occidental.* Instituto Pirenaico de Ecología. Jaca.

Penas, X. M & others (coords.) (1995). *Atlas de vertebrados de Galicia. Tomo II. Aves.* Consello da Cultura Galega. Santiago de Compostela.
Peris, S. J. & Carnero, J. I. (1988). *Atlas ornitológico de la provincia de Salamanca.* Diputación de Salamanca. Salamanca.
Picazo Talavera, J. (1989). *Aves de La Roda y Fuensanta (Albacete). Fenología, nidificación y comunidades.* Instituto de Estudios Albecetenses. Albacete.
Pleguezuelos, J. M. (1991). *Evolución histórica de la avifauna nidificante en el S. E. de la Península Ibérica (1850–1985).* Agencia de Medio Ambiente, Junta de Andalucía. Sevilla.
Pleguezuelos, J. M. (1992). *Avifauna nidificante de las Sierras Béticas Orientales y depresiones de Guadix, Baza y Granada: su cartografiado.* Universidad de Granada. Granada.
Pons, G. X. (ed.) (2000). *Las aves del Parque Nacional Marítimo-terrestre del Archipiélago de Cabrera (Islas Baleares, España).* Ministerio de Medio Ambiente y GOB. Palma de Mallorca.
Ramos, A. J. & Sarmiento, L. F. (1999). *Las aves de los humedales del sur de Alicante y su entorno.* Editorial Club Universitario. Alicante.
Rebassa, M., Muntaner, J. & Ramis, B. (2002). *Aus de les Illes Balears*. Perifèrics. Palma de Mallorca.
Ribas, J. (2000). *Els ocells del Vallès Oriental*. Lynx Edicions. Barcelona.
Rico, J. L. & Gil-Delgado, J. A. (1986). *Aves nidificantes en la provincia de Alicante.* Instituto Juan Gil-Albert, Diputación Provincial de Alicante. Alicante.
Román, J. & others (1996). *Atlas de las aves nidificantes de la provincia de Burgos.* Caja de Ahorros del Círculo Católico. Burgos.
Sáez-Royuela, R. (1990). *La guía de INCAFO de las aves de la Península Ibérica,* 2ª edición. INCAFO. Madrid.
Saiz, J. & Fombellida, I. (1999). *Aves de Cantabria.* Creática Ediciones. Santander.
Sampietro, F. J., Pelayo, E., Hernández, F. Cabrera, M. & Guiral, J. (eds.) (1998). *Aves de Aragón. Atlas de especies nidificantes.* Diputación General de Aragón and Ibercaja. Zaragoza.
San Segundo, C. (1990). *Atlas de las aves nidificantes de la provincia de Ávila y sierra de Gredos.* Universidad Complutense. Madrid.
Sánchez, T. (2002). *Aves de Canarias nidificantes*. Ed. Rueda. Alcorcón, Madrid.
Sandoval Rey, A. (2015). *Las aves marinas de Estaca de Bares.* Tundra Ediciones. Almenara (Castellón).
Sanz-Zuasti, J. & Velasco, T. (1997). *Guía de las aves de las lagunas de Villafáfila y su entorno.* ADRI "Palomares". Zamora.
Sanz-Zuasti, J. & Velasco, T. (1999). *Guía de las aves de Castilla y León.* Carlos Sánchez Alonso editor. Medina del Campo (Valladolid).
Sargatal, J. & del Hoyo, J. (1989). *Els ocells del Parc Natural dels Aiguamolls de l'Empordà.* Lynx Edicions. Barcelona.
SEO/Birdlife (2010). *Estado de conservación de las aves en España.* SEO/Birdlife. Madrid.
Snow, D. W. & Perrins, C. M. (eds.) (1998). *The Birds of the Western Palearctic, concise edition.* Oxford University Press. Oxford.
Tellería, J. L. (1981). *La migración de las aves en el Estrecho de Gibraltar. Volumen II. Aves no planeadoras.* Cátedra de Zoología de Vertebrados, Universidad Complutense. Madrid.
Tellería, J. L. (ed.). (1988). *Invernada de aves en la Península Ibérica.* Sociedad Española de Ornitología. Madrid.
Tellería, J. L., Asensio, B. & Díaz, M. (1999). *Aves ibéricas II. Paseriformes.* J. M. Reyero Editor. Madrid.
Tucker, G. M. & Heath, M. F. (1994). *Birds in Europe: their conservation status.* BirdLife International. Cambridge.
Urios, V., Escobar, J. V., Pardo, R. & Gómez, J. A. (1991). *Atlas de las aves nidificantes de la Comunidad Valenciana.* Conselleria d'Agricultura i Pesca, Generalitat Valenciana. Valencia.

Varela, J. M. (2007). *Aves amenazadas de España.* Lynx Edicions. Barcelona.
Villarino, A., González, S. & Bárcena, F. (2002). *Vertebrados da Limia, dende a lagoa de Antela ós nosos días. I. Aves: Gaviiformes a Piciformes*. Limaia. Ourense.
Woutersen, K. & Grasa, M. (2002). *Parque Nacional de Ordesa y Monte Perdido. Atlas de las aves*. Kees Woutersen Publicaciones. Huesca.
Woutersen, K. & Platteeuw, M. (1998). *Atlas de las aves de Huesca*. Ediciones La val de Onsera. Huesca.
Zuberogoitia, I. & Torres, J. J. (1998). *Aves acuáticas de Bizkaia.* Bilbao Bizkaia Kutxa. Bilbao.

IDENTIFICATION

Alström, P. & Mild, K. (2003). *Pipits & wagtails of Europe, Asia and North America.* Christopher Helm. London.
Alström, P., Colston, P. & Lewington, I. (1991). *A field guide to the rare birds of Britain and Europe.* HarperCollins Publishers. London.
Baker, K. (1993). *Identification guide to European non-passerines.* BTO Guide 24, British Trust for Ornithology. Thetford (Norfolk, Inglaterra).
Baker, K. (1997). *Warblers of Europe, Asia and North Africa.* Christopher Helm. London.
Beaman, M. & Madge, S. (1998). *The handbook of bird identifcation for Europe and the Western Palearctic.* Christopher Helm. London.
Byers, C., Olsson, U. & Curson, J. (1995). *Buntings and sparrows.* Pica Press. Mountfield (East Sussex, England).
Clement, P., Harris, A. & Davies, J. (1993). *Finches and Sparrows: an identification guide.* Christopher Helm. London.
Constantine, M. & The Sound Approach (2006). *The Sound Approach to birding. A guide to understanding bird song.* The Sound Approach. Poole (Dorset, England).
Forsman, D. (1999). *The raptors of Europe and the Middle East. A handbook of field identification.* T. & A.D. Poyser. London.
Génsbol, B. (2008). *Birds of prey.* Collins. London.
Grant, P. J. (1982). *Gulls. A guide to identification.* T. & A.D. Poyser. Calton (Staffordshire, England).
Harrison, P. (1983). *Seabirds, an identification guide.* Croom Helm. London.
Hayman, P., Marchant, J. & Prater, T. (1986). *Shorebirds. An identification guide to the waders of the world.* Croom Helm. London.
Jenni, L. & Winkler, R. (1994). *Moult and ageing of European Passerines. A photographic guide.* Academic Press. London.
Jonsson, L. (1992). *Birds of Europe, with North Africa and the Middle East.* Christopher Helm. London.
Jutglar, F. & Masó, A. (1999). *Aves de la Península Ibérica*. Planeta. Barcelona.
Kennerley, P. & Pearson, D. (2010). *Reed & bush warblers*. T. & A. D. Poyser. London.
Madge, S. & Burn, H. (1988). *Wildfowl. An identification guide to the ducks, geese and swans of the world.* Christopher Helm. London.
Malling Olsen, K. & Larsson, H. (1995). *Terns of Europe and North America.* Christopher Helm. London.
Malling Olsen, K. & Larsson, H. (1997). *Skuas and jaegers: a guide to the skuas and jaegers of the world.* Pica Press. Mountfield (East Sussex, England).
Malling Olsen, K. & Larsson, H. (2004). *Gulls of Europe, Asia and North America.* Christopher Helm. London.
Message, S. & Taylor, D. (2005). *Waders of Europe, Asia and North America.* Christopher Helm. London.
Olsson, U., Curson, J. & Byers, C. (1995). *Buntings and sparrows: A guide to the buntings and North American sparrows.* Pica Press. Mountfield (East Sussex, England).
Porter, R. F., Willis, I., Christensen, S. & Nielsen, B. P. (1981). *Flight identification of European raptors. 3rd edition.* T. & A.D. Poyser, Calton, England.
Shirihai, H., Gargallo, G. & Helbig, A. (2001). *Sylvia Warblers.* Christopher Helm. London.

Svensson, L. (1992). *Identification guide to European passerines.* 4th edition. Stockholm, Sweden.
Svensson, L. & others (2010). *Collins bird guide.* 2nd edition. Collins. London.
Van Duivendijk, N. (2010). *Advanced Bird ID Guide.* New Holland. London.
Vinicombe, K., Harris, A. & Tucker, L. (2014). *The Helm guide to bird identification.* Bloomsbury. London.

WHERE TO WATCH BIRDS

Butler, J. R. (2001). *Birdwatching on Spain's southern coast*. Eds. Santana. Fuengirola (Málaga).
Chiclana, F. & Garzón, J. (2006). *Donde ver aves en Doñana.* Lynx Edicions. Barcelona.
Collins, D. & Clarke, A. G. (1996). *A birdwatcher's guide to the Canary Islands.* Prion. Perry (Cambridgeshire, England).
De Juana, E. (coord.). (1994). *Dónde ver aves en España Peninsular.* Sociedad Española de Ornitología and Lynx Edicions. Barcelona.
Del Hoyo, J. & Sargatal, J. (1989). *On observar ocells a Catalunya.* Lynx Edicions. Barcelona.
Garcia, E. & Rebane, M. (2017). *Where to watch birds in northern and eastern Spain.* 3rd edition. Christopher Helm. London.
Garcia, E. & Paterson, A. (2008). *Where to watch birds in southern and western Spain: Andalucia, Extremadura and Gibraltar.* 3rd edition. Christopher Helm. London.
Garcia, E. & Paterson, A. M. (1995). *Dónde observar aves en la España meridional.* Omega. Barcelona.
García del Rey, E. (2000). *Dónde ver aves en Tenerife*. Turquesa Publicaciones. Tenerife.
Hearl, G. (1999). *A birdwatching guide to Menorca, Ibiza and Formentera.* Arlequin Press. Chelmsford (Essex, England).
Hearl, G. (1999). *A birdwatching guide to Mallorca.* Arlequin Press. Chelmsford (Essesx, England).
López Gallego, A. (2000). *Dónde ver aves y naturaleza en Extremadura*. Ed. Albaragena. Monesterio (Badajoz).
Montero, J. A. & SEO/BirdLife (2005). *Where to watch birds in Spain. The 100 best sites*. Lynx Edicions. Barcelona.
West, S. (2004). *Where the birds are in northeast Spain.* Privately published (westvidal@teleline.es).

JOURNALS

Animal Biodiversity and Conservation (formerly *Miscel.lània Zoològica*). Museu de Zoologia, Barcelona. http://abc.museucienciesjournals.cat/?lang=es
Ardeola. Sociedad Española de Ornitología, Madrid. http://www.ardeola.org/
Aves & Naturaleza (formerly *La Garcilla*). SEO/BirdLife, Madrid. http://www.seo.org/category/aves-y-naturaleza/
Ecología. Organismo Autónomo Parques Nacionales, Madrid. http://www.magrama.gob.es/es/ministerio/servicios/publicaciones/rev_numero_art.asp?codrevista=REPN
Revista Catalana d'Ornitologia (formerly *Butlletí del Grup Català d'Anellament*). Institut Català d'Ornitologia, Barcelona. http://revistacatalanaornitologia.blogspot.com.es/

REGIONAL BIRD REPORTS

Albacete http://anuario.albacete.org/
Aragón http://anuariorocin.blogspot.com.es/
Asturias http://www.coa.org.es/index.php/home
Baleares http://www.gobmallorca.com/informa-t/publicacions/content/355-anuari-ornitologic
Burgos http://www.avesdeburgos.com/anuario.htm

Cádiz http://cocn.tarifainfo.com/anuariocadiz/inicioanuario.
Cantabria http://aves.eldelweb.com/Cantabria/noticias
Ciudad Real http://seocranuario.blogspot.com.es/
Comunidad Valenciana http://www.internatura.org/aocv/
Extremadura http://www.adenex.org/index.php?option=com_content&task=view&id=117
Galicia http://sgo.cesga.es/aves/htdocs/
Gibraltar http://www.gonhs.org/publications.htm
Granada http://www.aefona.org/anuario-ornitologico-de-la-provincia-de-granada/
Guipúzcoa http://gorkagorospe.com/portfolio-item/anuario-ornitologico-de-gipuzkoa/
La Rioja http://www.ecologistasenaccion.org/article4940.html
Madrid (1996–2008). http://www.seomonticola.org/category/anuario-ornitologico-de-madrid/
Málaga http://www.seomalaga.org/index.php?menu=anuario
Murcia http://aorm.blogspot.com.es/
Navarra https://sites.google.com/site/scngorosti/Home/secciones/seccion-de-zoologia/ornitologia/anuario-ornitologico-de-navarra
Palencia http://www.avespalencia.org/index.php/anuario-ornitologico-de-palencia
Sevilla http://gosursevilla.blogspot.com.es/p/publicaciones.html
Toledo http://esparvel.org/anuario/anuario-ornitologico-toledo-normas-colaboracion/

OTHER WEBSITES OF INTEREST

Atlas de las aves reproductoras de España http://www.magrama.gob.es/es/biodiversidad/temas/inventarios-nacionales/inventario-especies-terrestres/inventario-nacional-de-biodiversidad/ieet_aves_atlas.aspx
Atlas de las aves en invierno en España http://www.magrama.gob.es/es/biodiversidad/publicaciones/atlas_aves_invierno_tcm7-291664.pdf
Aves de Ceuta http://www.avesdeceuta.com/
Aves de Extremadura http://aves-extremadura.blogspot.com.es/
Aves Ibéricas (Miguel Rouco) http://www.avesibericas.es/links.htm
Avibase – the world bird database http://avibase.bsc-eoc.org/avibase.jsp?lang=EN
Canary Islands Birding News http://avescanarias.blogspot.com.es/
Comité de Rarezas de SEO/BirdLife http://www.seo.org/conocenos/grupos-de-trabajo/comite-de-rarezas/
Enciclopedia de las aves de España (SEO/BirdLife) http://www.seo.org/listado-aves/
Enciclopedia Virtual de los Vertebrados Españoles (Museo Nacional de Ciencias Naturales, Madrid) http://www.vertebradosibericos.org/
Fatbirder http://fatbirder.com/
Handbook of the Birds of the World Alive (Lynx Edicions, Barcelona) http://www.hbw.com/
Internet Bird Collection (IBC) http://ibc.lynxeds.com/
La migración de las aves (SEO/BirdLife) http://www.migraciondeaves.org/
Libro rojo de las aves en España http://www.magrama.gob.es/es/biodiversidad/temas/inventarios-nacionales/LR_completo_para_web_tcm7-164856.pdf
Rare Birds in Spain http://www.rarebirdspain.net/home.htm
Reservoir Birds http://www.reservoirbirds.com/
Surfbirds – the world birding web site http://www.surfbirds.com/
Xeno-canto (cantos de las aves de Europa) http://www.xeno-canto.org/collection/species/all?area=europe

USEFUL ADDRESSES

SEO/BirdLife
C/. Melquiades Biencinto, 34
28053 Madrid
Tel. 91 4340910
E-mail: seo@seo.org
Web: http://www.seo.org/

Lynx Edicions
C/. Montseny, 8
08193 Bellaterra (Barcelona)
Tel. 93 5947710
E-mail:
Web: http://www.lynxeds.com

BirdLife International
The David Attenborough Building, Pembroke Street
Cambridge CB2 3QZ (UK)
Web: http://www.birdlife.org/

Others

AERC (Association of European rarities committees) http://www.aerc.eu/
Alosa – Sonidos de la naturaleza http://www.sonidosdelanaturaleza.com/
Biodiversidad (MAGRAMA) http://www.magrama.gob.es/es/biodiversidad/temas/
Euring (scientific ringing) http://www.euring.org/
Gibraltar Ornithological and Natural History Society (Birdlife in Gibraltar) http://www.gonhs.org/
Grup Balear d'Ornitologia i Defensa de la Naturalesa http://www.gobmallorca.com/
Institut Català d'Ornitologia http://www.ornitologia.org/ca/
Óptica Roma http://opticaroma.com/
Oryx (La tienda del amante de la naturaleza) https://www.weboryx.com/oryx/
RSPB (BirdLife in UK) http://www.rspb.org.uk/
Servicio de Protección de la Naturaleza (SEPRONA, Guardia Civil) http://www.seprona.es/
SPEA (BirdLife in Portugal) http://www.spea.pt/pt/

SPECIES INDEX

English name (in bold type)
Scientific name (in oblique type)
Spanish name (in roman type)

D

T

BIRD TOPOGRAPHY

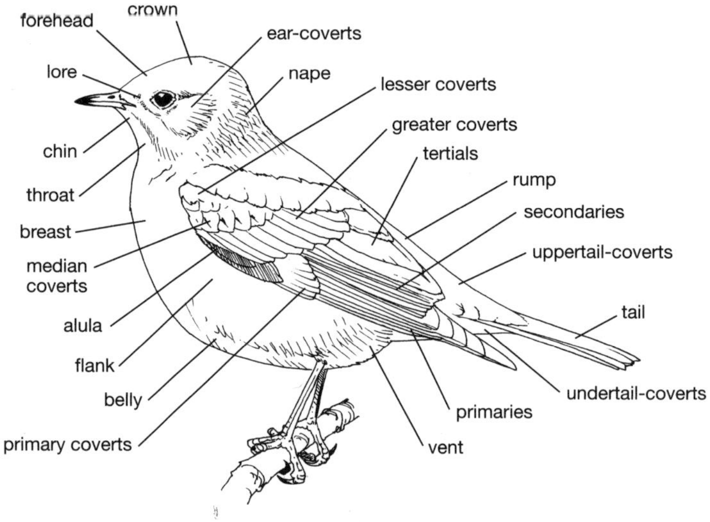

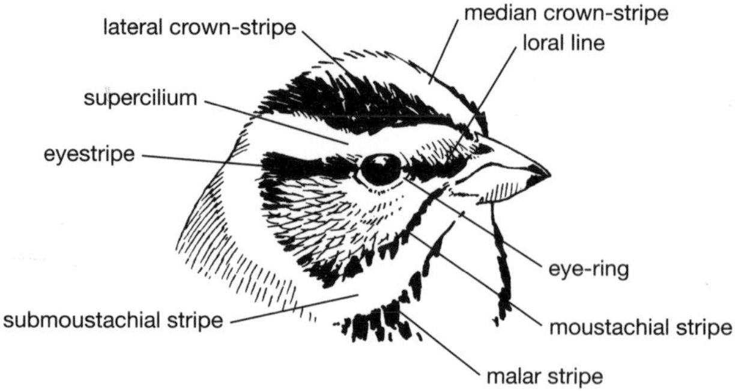

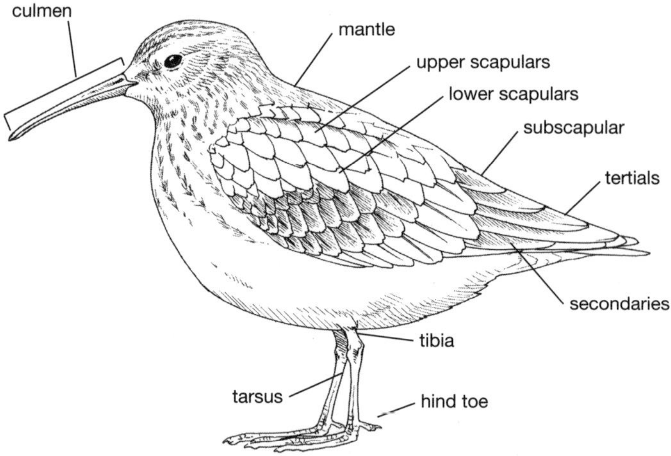